Perspectives on Defense Systems Analysis

The What, the Why, and the Who, but Mostly the How of Broad Defense Systems Analysis

MIT Lincoln Laboratory Series

Perspectives on Defense Systems Analysis: The What, the Why, and the Who, but Mostly the How of Broad Defense Systems Analysis, William P. Delaney

Ultrawideband Phased Array Antenna Technology for Sensing and Communications Systems, Alan J. Fenn and Peter T. Hurst

Decision Making Under Uncertainty: Theory and Application, Mykel J. Kochenderfer

Applied State Estimation and Correlation, Chaw-Bing Chang and Keh-Ping Dunn

MIT Lincoln Laboratory is a federally funded research and development center that applies advanced technology to problems of national security. The books in the *MIT Lincoln Laboratory Series* cover a broad range of technology areas in which Lincoln Laboratory has made leading contributions. This series renews the knowledge-sharing tradition established by the seminal *MIT Radiation Laboratory Series* published between 1947 and 1953.

Perspectives on Defense Systems Analysis

The What, the Why, and the Who, but Mostly the How of Broad Defense Systems Analysis

William P. Delaney

with

Robert G. Atkins
Alan D. Bernard
Don M. Boroson
David J. Ebel
Aryeh Feder
Jack G. Fleischman
Michael P. Shatz
Robert Stein
Stephen D. Weiner

The MIT Press
Cambridge, Massachusetts
London, England

This book was set in Garamond by MIT Lincoln Laboratory, Lexington, Massachusetts.

Library of Congress Cataloging-in-Publication Data is available.

ISBN: 978-0-262-02935-3 (hardcover)
ISBN: 978-0-262-54972-1 (paperback)

Dedication

Dedicated to defense systems analysts everywhere. They bring a degree of clarity, insight, and coherence to complex technical problems and questions in national security. We particularly remember one of the very best, Dennis Murray.

Table of Contents

Acknowledgments

The incentive for this book came from Dr. Eric Evans, Director of MIT Lincoln Laboratory. Eric came into my office late in 2010 and told me I should write a book on defense systems analysis. Shortly thereafter, he initiated a new Lincoln Laboratory book series with this book in mind as one of the first in that series.

While I had participated in many systems analyses over the course of my career at the Laboratory, I did not feel that I alone had enough insights for a book, though Eric was confident that I did. I realized that Lincoln Laboratory has many talented systems analysts, most of whom developed their skills in programs I created or led. Some arm-twisting ensued to gather chapters from these busy folks, but their contributions have made this whole enterprise worthy of a book. My thanks to them, and in particular to Dr. Evans for his initial strong push.

The authors acknowledge and thank Margarida Santos, Stephanie Mosely, Libby Samarjian, and Dorothy Ryan for their substantial contributions to the drafting, editing, and publication of this manuscript.

Foreword

> This is as strange a maze as e'er men trod... Some oracle
> Must rectify our knowledge.
>
> —William Shakespeare, *The Tempest*

The Department of Defense and its military components are a continuing source of complex scientific, engineering, and technological problems. The Department is often in the role of "pushing the envelope" on new techniques and systems or taking on hitherto unsolved technical issues in the interest of national security.

These problems can be poorly described, vaguely understood, or otherwise not "well set." The closest thing we have to Shakespeare's oracle above are some folks who have a penchant for addressing complex, ill-stated, and uncertain problems and analyzing their way through the maze to a clearer statement of the main issues, the cause and effect relationships, and the major uncertainties. Often they can point the way to the solution.

I call these explorers systems analysts, and I believe our authors are such explorers. They represent more than 250 years of experience and background in this art. And it is a bit of an art; it is not specifically taught in engineering and science curriculums, and its practitioners come from a variety of disciplines. We will discuss this art of systems analysis in detail, but first, what is the purpose of our modest book?

We hope to help new or current systems analysts with some insights from our experience, and we further hope to convince some young engineers and scientists that they can become practitioners of this art. A technical organization doesn't need a lot of systems analysts, but they surely need a few, and we hope this writing can serve as a primer to the art and an encouragement to some standing on the sidelines—afraid to get their feet wet!

Bill Delaney

At the start
January 2011
Longboat Key, Florida

Introduction to the Authors

William P. Delaney

Most of our authors are from MIT Lincoln Laboratory, where we have done many broad defense systems analyses (and we like to think we do some of the best systems analysis around). Robert Stein, formerly Vice President of Raytheon, joins us with his view of this type of work in the aerospace and electronics industry.

Each chapter has a short biography of the author at the end. Most authors also at some point provide an informal rendition of their introduction into the systems analysis tradecraft.

I

Systems Analysis Overview

1

The Goal and Organization of This Book

William P. Delaney

Defense systems analysis is not a new art, and it certainly was not originated by MIT Lincoln Laboratory. In fact, Lincoln Laboratory owes its origin to a systems analysis, which is explained in Chapter 4. Before the Laboratory's founding in the early 1950s, systems analysis was used in World War II by British operations researchers to explore how to reduce attrition of their bomber aircraft making sorties over Europe. In addition, in response to German submarines' sinking 3000 Allied ships in the North Atlantic on their way to Britain, a team of British and U.S. analysts deduced how to counter these submarines. The insight of these analysts resulted in the sudden end to the German submarine threat, as more than 700 submarines were sunk in less than two years!

In the 1960s, systems analysis became a buzzword in the Kennedy administration. Secretary of Defense Robert McNamara and his cadre of "whiz kids" led by Alain Enthoven (who later was appointed Assistant Secretary of Defense for Systems Analysis) invented a class of mission analysis that changed the department's acquisition process. In the late 1960s, the RAND corporation published a number of books on this application of systems analysis to policy decisions on DoD weapon systems.[1] During this era, Lieutenant General Glenn Kent, U.S. Air Force, pioneered the application of systems analysis to the Air Force's major weapon systems questions.[2] Throughout the Cold War, systems analysis underwrote our strategic posture of deterrence, and systems analysis was widely used in the intense debates surrounding

1 See for example, *Systems Analysis and Policy Planning: Applications in Defense*, E.S. Quade and W.I. Boucher, eds. New York: American Elsevier, 1968.

2 Glenn A. Kent, *Thinking About America's Defense: An Analytical Memoir*. Santa Monica, Calif.: RAND Corporation, 2008.

national missile defense. Our authors have carried on this analysis tradition, applying the techniques to a widening variety of defense issues.

The goal of this book is to provide insight, advice, and encouragement to those involved (or thinking about getting involved) in the analysis of defense systems.

We lean toward the newcomers in our potential audience. We believe our advice is useful for all types of broad systems analysis beyond just defense systems, and some of us have used these insights in studies for entities such as the Federal Aviation Administration and the Department of Homeland Security.

My first three chapters dissect our kind of systems analysis, explaining what it is, why one wants to do the process, and mainly how one might get it done. Since the process is a bit of an art, there is no fixed formula to follow for doing it, and different practitioners have varying viewpoints and advice to give. I have assembled here the insights of nine other experienced practitioners to provide a relatively broad spectrum of viewpoints.

Since most of the contributing authors have initially trained in this art in a few major program areas at Lincoln Laboratory, the reader will see a number of oft-repeated themes (get the question right, tell a good story, check and recheck all inputs, explain all your assumptions, eschew details, etc.). I have elected to let this redundancy remain as a reinforcement of important elements of advice.

We all may radiate a degree of self-confidence since we think we are pretty good at this analysis process, but I hope we do not sound self-congratulatory or arrogant. The humbling thought is, we have probably made all of the mistakes cautioned against in this writing!

I define just what type of analysis is our focus. *Systems analysis* means many things to many people. If you go to a well-stocked technical library and draw out all books with systems analysis in the title, you will have a heavy stack of books. So we need a degree of specificity in our particular use of the term. Why one does this kind of analysis also needs to be addressed, and I argue the case for doing it is simple: it is a valuable element of the problem-solving process. It is not the whole process but a key part of the "getting started" drill.

I also explore the various ways one may migrate into this broad defense systems analysis business. Generally one does not start off with a plan to become such an analyst, but then something happens along the way.

My views on the "how" of this kind of analysis constitute most of Chapter 3. There are different venues that need to be discussed, such as studies internal to one's organization and studies conducted for outside entities such as the various advisory boards of the military or the Department of Defense (DoD) (for example, the Air Force or Army Scientific Advisory Boards, the Naval Research Advisory Committee, and the Defense

Science Board). Different venues have different ways of operating, and I discuss them separately although much of the advice applies to both. I emphasize external work since I spent much of my last 20 years in that venue. My contributing authors mostly discuss internal study venues, which may be of higher interest to newcomers.

Chapter 3 provides practical advice on how to set up and run a study. The remarks are directed to the potential study leader or study committee chair. This leader role is challenging, but it is the single most important determinant of success in the whole study enterprise. In my experience, I coveted this role if I had a background in the topic or the right background to take on a relatively unexplored area.

Examples, synopses, and "war stories" of a variety of systems analysis efforts from my experience are sprinkled throughout my writing and the writings of our contributing authors. My examples are selected to illustrate the variety of such efforts and to point to the sometimes substantial impact of such studies, the often delayed impact, and in some cases the lack of impact. Chapter 4 is a compilation of examples from experiences when most of the time I was the chairperson of the study.

Most of us are limited in the level of detail we present in our examples because most defense systems analyses involve classified information. Occasionally, we can present detail since the issue being studied did not involve classification (e.g., Chapter 13 by Don Boroson and Dave Ebel's wind-turbine case study in Chapter 9).

My contributing authors present their perspectives on this analysis process in subsequent chapters. All of them have led teams of analysts and have introduced many newcomers to the systems analysis tradecraft. They were free to cover the ground in their own format and style, and provide a view independent of the advice offered elsewhere in the book. They have worked in somewhat different technical areas, so their insights broaden the advice we have to offer.

They have chosen different paths to make their case. Some like Bob Stein, Aryeh Feder, Bob Atkins, and Alan Bernard offer a broad or personalized rendition of their experience in the systems analysis game, and I have grouped their chapters together after my introductory chapters. The other authors—Dave Ebel, Steve Weiner, Michael Shatz, Don Boroson, and Jack Fleischman— stick quite close to their areas of technical expertise and dissect the systems analysis protocols of those areas. Their chapters make up the later part of this book.

I call this approach to composing a book the "shotgun" approach, in which one has a number of experienced voices give their perspectives on the topic of interest. Imagine a book entitled *Beginning Golf* by 10 well-known golf pros; the book may not assure a formula for success, but some chapters will resonate with some golfers and some common themes will emerge, and that is often the best authors can hope for. Our authors' voices are those of experienced (read *older*) investigators; there are no youngsters among us, but this systems analysis tradecraft we profess has numerous timeless features that we offer for our readers.

I return in the final short chapter to summarize prominent themes and to address a question that will surely be on the reader's mind: "Is this systems analysis stuff for me?"

Hopefully, we provide some pearls of wisdom along the way.

2

The What, Why, and Who of Defense Systems Analysis

William P. Delaney

> Where am I? How did I get here? Where do I want to go?
> How am I going to get there?

2.1 The What

The above guide to the student of feedback control theory was a favorite of one of my undergraduate professors in the late 1950s. It is also a good definition of the type of broad systems analysis that is our focus here.

Our type of analysis is often the first (or an early) comprehensive and coherent dissection of the problem space conducted by analysts with the appropriate scientific credentials. The analysis is broad to encompass all the issues, and it is based on the laws of physics, mathematics, and the engineering disciplines.

The analysis may or may not delineate a solution to the problem, but it should provide the follow-on workers a substantial roadmap of the issues they face. It should uncover the major cause-and-effect relationships, and it may suggest likely solution options or paths. It may be able to provide a rough estimate of the costs of solving the problem. It should uncover unsolved phenomenological issues and other types of "showstoppers." This systems analysis process ends with an articulation of the problem in a clear, digestible fashion, and promising solution paths are presented in a quantitative manner. It is all done quickly (weeks, months, rarely a year).

A thorough systems analysis will occasionally discourage the government from taking on the issue when they realize the roadblocks or the extreme costs likely in the enterprise.

Thus, our systems analysts function much as explorers, or "the first folks ashore." They have a touch of the hero since they enter in where others have not tread (or coherently tread). They are a powerful force for getting the solution right at the start by their delineation of the issues and the possible solution paths. I will say more about this explorer role later.

Key descriptions of this particular brand of systems analysis are broad, comprehensive, science-based, done early and rather quickly (weeks or months, not years). Outputs are briefings and a written report.

The type of analyses we represent is not engineering design studies, data processing architectural studies, analyses of quite specific scientific issues, broad economic studies, management consulting efforts, proposal preparations, or a wide variety of other activities that have "system" in their title or description.

2.1.1 Examples of Our Type of Systems Analysis

1950s	How can the United States defend against Soviet bombers carrying nuclear weapons (resulted in the SAGE continental air defense system, and the impetus for the formation of MIT Lincoln Laboratory and later the MITRE Corporation; see Chapter 4)?
1955	How do we detect intercontinental ballistic missiles approaching the United States (resulted in the BMEWS system)?
1960s	How can we defend urban United States against Soviet missile attack?
1970s	What are the comparative survivabilities of the B-1 bomber and the modern cruise missile in the strategic penetration of Soviet air defenses (see Chapter 4)?
1980s	What are the options for active defense of Minuteman missile silos against Soviet missile attack?
	How do we defend the United States against air attack by cruise missiles?
1980s to Present	What are the vulnerabilities of U.S. military air vehicles (including low-observable vehicles) against all types of air defense, both conventional and unconventional (see Chapter 6)?
1990s	How vulnerable is the GPS system to electronic countermeasures? What are the options for reinforcing the system (see Chapter 4)?
2005	What are the options for removing unexploded munitions from our domestic training ranges (see Chapter 4)?
Present	What are the potential solutions or counters to improvised explosive devices? What are the prospects for defense against biological attacks (see Chapter 11)?

By now the reader has an idea of our type of analysis. I next present arguments for doing this type of analysis, but I suspect I have already given away the key arguments by the descriptions above.

2.2 The Why

Simply stated, this type of analysis is the classical engineering approach to a new and challenging problem area. The analysis reviews what has been done in the past and what are the circumstances that cause the problem today. The main part of the analysis is a walk-through of the total problem space. It provides a measure of the scale and the difficulty of the challenge. Major uncertainties and phenomenological issues are identified; key investigators in the field are found, and their work and results are assimilated. The rough scale of funding involved in tackling this challenge can often be estimated.

An important result of this analysis is a clear, logical, and technically informed articulation of the above facets of the problem. Decision makers can then determine if they want to proceed, and this analysis will become the front end of their proposal to do so. The clear articulation of the issues in scientific and engineering terminology will encourage competent analysts and investigators to join any follow-on effort to solve the problem.

The above rendition leaves out a lot of the real world. In the real world, there is often a high level of confusion about the nature of the problem. There can be ill-informed speculation on the causes of the problem, and there are often plentiful offers of solutions by the uninformed or technically inept folks with something to sell and by others with axes of one type or another to grind. The analysis we advocate attacks this chaotic situation by applying the laws of science and engineering (and logic) to identify cause-and-effect relationships and generally sort out the mess so that some degree of order can prevail. Scientists and engineers are right at home doing this if given the chance.

This degree of order will serve to chase out the lightweights, the uninformed, and the dilettantes. Further, this degree of order is necessary to capture any substantial government funding for a serious follow-on effort on the problem. The group that can provide a degree of order out of the chaos will be in a position to participate in the follow-on effort to the extent that they want to be continuing players.

2.3 The Who

My experience suggests that nobody starts out to be a systems analyst; folks get there by a variety of paths, and our authors tell their own stories. One may join a group of analysts and just start doing what they are doing. I know contributing author Steve Weiner, trained as a physicist, joined a small group doing broad analysis in missile defense, and quickly became very good at it. Alternately, one may be involved in a major project and questions arise as to the bigger picture in which this project is enmeshed, and these questions fall to the individual to assess and answer.

Contributing author Bob Stein suggests he got there by not being good at electronic circuit design, but I have a more plausible explanation. The systems analysis talent is a bit rare and engineering teams badly need that talent, so if you show some glimmer of it, as did Bob Stein, you will quickly be drafted to fill that role.

My own experience was a bit like Bob Stein's. I was working on early phased array radar technology in a pioneering group at Lincoln Laboratory. I wound up doing a lot of antenna experimentation since I was not very good at the underlying complicated electromagnetic theory and (to me) unwieldy mathematics.

Phased arrays promised to be very expensive, and the only defense project with the extreme need and deep enough pockets to consider them was ballistic missile defense. Missile defense in the early 1960s was somewhat tumultuous; the government changed its mind every few years as to the missile defense role, and each change was followed by a plethora of studies by analysis groups around the nation as to the architecture and technology of the new system. So I migrated (a bit reluctantly at first) into missile defense systems analysis, following in the footsteps of a close friend and colleague, Joel Resnick, who was quickly becoming Lincoln Laboratory's leading analytic force in missile defense analysis. I regard Joel Resnick as the prototype of the Lincoln Laboratory defense systems analyst. He was very bright with quick insight into a complex issue and could ferret out the dominant issues better than anyone else. He had an unpredictable briefing style that kept everybody interested. His chaotic work style left a lot of analysis needs lying around for me to grab. We were later joined by John Fielding, another "natural" for broad defense analysis. Our division head, Jerry Freedman, was an outspoken expert in air defense who wanted to become expert in missile defense, so he was an avid listener. We formed an exciting (noisy) group, and along the way, I developed some penchant for asking and sometimes answering key questions about missile defense architectures. This missile defense analysis tradition has been carried forward for many years in a much quieter and more competent manner by contributing author Steve Weiner.

This early era in missile defense proved to be an exciting field, populated with very bright folks from industry, Bell Laboratories, defense laboratories, think tanks, and what I called the "physics factories" of Los Alamos, Lawrence Livermore, and Sandia. We were high-visibility folks since we were on the front edge of big system postulations, and the whole missile defense field was very controversial. We were youngish folks (in our early 30s), but we found ourselves arguing our particular findings in front of high-ranking Department of Defense executives and nationally prominent scientists. My Laboratory higher-ups, particularly Jerry Freedman, were very interested in our work, and we did lots of dry runs for them before they let us off to Washington.

War Story

Somewhere around 1967, I recall giving a talk on high-altitude nuclear blackout effects on missile defense radars to the President's Scientific Advisory Committee, PSAC. Nobel Prize–winning physicist Professor Hans Bethe of Cornell University was on the committee, and he surely had forgotten more physics than I would ever know.

The physics of high-altitude blackout was extremely uncertain, so I had taken the somewhat desperate approach of corralling all the knowledgeable nuclear scientists on this topic at the University of California for three days and essentially holding an auction to establish the coefficients we needed to calculate the blackout. We started at low altitudes and smaller nuclear yields where things were most certain and then migrated upwards in altitude and weapon yields to where things got decidedly squirrelly.

Predictably, at the end of my talk, the kindly professor asked me, "Mr. Delaney, what high-altitude chemistry and physics did you use to predict these results?" With my heart in my mouth, I said, "Professor Bethe, there is very little chemistry or physics in my calculations," and I described my auction process. Bethe sort of chortled and said, "That is the only answer to give, and, by the way, we did a lot of that at Los Alamos during the Manhattan Project." So these were exciting times for a young engineer.

The attractions then of this systems analysis role are the satisfaction of exploring a complex but important problem, the visibility that comes from being on the leading edge of a critical issue, and the chance to work myriad different and challenging projects with some of the best and finest in the nation.

In my case, this attraction was strengthened by my realization that I did not have the talent to be a phased array radar expert, but seemed to have some "feel" for this system stuff.

I want to return to the explorer role of the systems analyst. The needed skills and personality traits include exploring a space without getting lost down one hole; holding seemingly contradictory facts in tension to see if the exploration resolves the apparent conflict; being comfortable and patient in a cloud of uncertainty; simultaneously assimilating; and remaining focused on stitching together the coherent picture while constantly challenging conclusions. In a nutshell, one needs abstract thinking skills, a capacity to assimilate and cohere a picture, and a lot of patience and perseverance. With practice, one learns to do this in a sprint, not a slow walk! It is not as daunting as it sounds.

My boss used to ask me, "What makes a good systems analyst, and how do we find them?" There is no academic discipline that trains such folks, and I would be suspicious of some academic regime that claimed it did! However, in many scientific or engineering projects, there is a limited cadre of folks who start to ask, "Tell me again why we are doing this or why are we doing it this way." They have maybe raised their heads up from an intense "deep dive" on some topic and have begun to question the broader aspects of the problem, asking why and how. I would encourage their questioning and start to give them broader systems analysis tasks.

A relevant question is, "Are broad system analysts born or created?" I don't know the answer to this question; I guess it is both paths. I have come across what I call "naturals" in my career. I could see the intellectual leaning to the bigger questions and also the personality traits to do this type of analysis and "sell" the story to senior leadership. The individual may not realize they have "got it," but I could see it.

Most eventual broad thinkers come with some scientific strength or background that acts as a springboard to the broader questions of how their "stuff" gets used. My personal favorite hunting ground for converts was nuclear physicists; they had great training in physics and mathematics, and the issues they took on in their graduate work were often very broad in concept. I saw them as relatively fearless in dealing with complex physical phenomena and the associated hairy mathematics. Too bad they could not find jobs doing nuclear physics!

Another (Short) War Story

In the early 1970s, times were tough for physicists, and I interviewed this physicist who appeared to be extremely bright and also had a very good feel for hardware issues. I recall telling him we were searching for broad system guys and asking if he would be interested in that type of analysis. His comeback: "I did my PhD on the origin of the universe—how broad do you want to get?" He was one of the best persons I ever hired.

The bottom line on broad systems analysts is, you can run across some in your activities, but many times you sense the promising raw material and "grow" them. It is a satisfying process, but you have to become adept at convincing them that this style of work is legitimate, interesting, important, and career-building. In essence, you are weaning them away from their vision of becoming a national expert in some technology or a Nobel Laureate in physics!

3

The How of Defense Systems Analysis

William P. Delaney

There is no formula for how to conduct these broad defense systems analyses. This book provides a range of perspectives from a variety of practitioners on how to go about these analyses. These early chapters convey my insights, and the nine contributing authors to follow provide theirs. Forgive my tendency to lapse into occasional "war stories," but they do convey insights sometimes better than words of advice.

From my experience, I see two distinct situations for the conduct of such analyses. One is a study that is done largely within one's organization. The incentive for the study may be internally generated, or it may come from a sponsor's request. Most of the staffing for the study will come from inside one's organization, and the methods and protocols for conducting the study are largely internal choices. Usually, you have a pretty solid understanding of the question or issue to be addressed in this class of study. The study can be approached informally and set up with a lot of flexibility. You can plot your own course guided by technical judgment and scientific intuition. Your team members are generally well known to you and are well motivated to be team players for the benefit of your organization. This class of systems analysis is a good training ground for honing your skills in finding a path through a complex issue and developing skill at presenting your group's insights to senior management.

The second study type is one done in more formal circumstances, such as task forces conducted by various advisory boards to the military or other DoD leadership. Examples here are studies of the Defense Science Board (DSB) and the Air Force Scientific Advisory Board. I have served on both, with my main involvement being with the DSB. I have also served on many other such government panels, often treating classified topics.

These studies are a decidedly different breed. There is more formality and more concern for openness and a balanced team without conflict of interest. You often start with more diverse positions and insights that will need to be reconciled as you proceed. The real question to be addressed may take some time to discover. You will have much information pushed on you, and this information may have significant "spin" and self-protection of some program or agency built in. My advice for these studies is going to be quite a bit more specific, maybe even a bit dogmatic. I start with the easier case of internal studies, reiterating that much of this advice is common to both types of study.

3.1 Internal Broad Systems Analyses

I got involved in this type of analysis early in my career in the 1960s. The topics revolved mostly around radar applications to ballistic missile defense. Missile defense of that era was even more controversial than it is now, and the nation's objectives for missile defense seemed to change frequently. Thus, there was much scurrying about on issues such as urban defense against heavy attacks, defense against accidental or light attacks, defense of our missile silos, etc. I seemed to be involved in a continuous sequence of such issues. We were a small group at Lincoln Laboratory, and there were other groups at think tanks, a few industrial firms, and most importantly the renowned Bell Telephone Laboratories, who had the major R&D responsibility for the Army's national missile defense mission.

Some of the issues were architectural in nature, such as how to deploy one's radars and interceptor missiles to defend an ensemble of targets. Many analyses focused on the possible countermeasure aspects of an enemy's attack, and this debate was fueled by an aggressive program within the U.S. Air Force ballistic missile offense community to develop all kinds of countermeasure devices for their missile systems. It really was an "us battling us" type era that kept systems analysts around the country very busy. We were also fueled by data from almost weekly full-scale tests at Kwajalein Atoll, the nation's major test site.

This era was a great training ground for systems analysts, and missile defense continues to be a rich source of this type of analysis (author Steve Weiner has had an interesting career in it for close to 50 years!). Part of the valuable experience and training from these internal studies came from the group of interested and technically very astute listeners in the DoD and the military. They brought a degree of focus and energy that encouraged us to be at the top of our game when we briefed them, often with results at some variance with those being presented by Bell Laboratories. (Some of my most admired recipients were Dan Fink and Major General John Toomay of the Office of the Secretary of Defense and Major General Jasper Welch of the Air Force. John Foster, Gene Fubini, Julian Davidson, Bert Fowler, Dave Heebner,

Ben Alexander, Seymour Zieberg, Sam Rabinowitz, and Jerry Freedman were also among the great listeners). These names are largely unknown to our younger readers. Many of these giants have passed on, but I mention them to salute their important intellectual leadership throughout the Cold War. These technically astute Cold War warriors didn't just listen quietly; they were questioning, outspoken, and critical, and those traits helped us develop a focus on the real issues and an ability to articulate them.

3.1.1 The Role of Chairperson

My advice on both internal and external systems analysis is directed to potential leaders or chairpersons of both internal and external studies. My number one advice is that the selection of a great chairperson is the single most important determinant of the success of a systems analysis enterprise. An internal study can possibly do without a great chair as long as it has a very solid set of analysts, but the further outside the organization your study reaches, the more you need "the best."

So if you are asked to be the chairperson, be flattered but be aware that the role is demanding. Along with the principal leadership role, you have to play scoutmaster to a sometimes unruly band. Really good systems analysts and supporting scientific gurus are generally an independent bunch, and somehow you have to keep them interested and focused. I have found that the higher-up person in the organization or the important sponsor who has initiated the internal study plays an important role. Their presence at the start and at periodic reviews is an important focusing tool. I recall my words from a study in the 1970s: "Come on you guys, quit fooling around! What are we going to tell our tough boss Jerry Freedman on Friday?"

You intellectually lead the whole drill by continually posing the main questions and expressing your own confusion with what to conclude or your inability to understand the more obscure analyses. You can play the role of a dummy and challenge the group to paint a logical path for you through the whole morass of information. George Heilmeier of the DSB (former Director of DARPA) was fond of challenging his groups late everyday with, "If right now I had to tell the undersecretary what we are concluding, what do I say?" It is a good drill to follow.

Along with the scoutmaster and intellectual lead roles is the ombudsman role. You have to look out for your team and protect them from time-wasting briefings and diverting activities. You should make an obvious point of treating the group's time together as a precious commodity, and if some speaker is droning on and on, you need to be the tough guy and move that speaker off. You worry about the comfort of the study facility, the temperature of the coffee, and the morale of the members. In a good study, nobody will be happy, but your one-on-one short dialogs with individual members keep them content and keep you in tune with the undercurrents that can affect the progress or morale of the whole enterprise.

The informality of the internal study gives one a lot of latitude in operation. One problem is attracting the best people and keeping them involved despite the responsibilities of their day jobs. I was spoiled in my early career because I worked mainly with a group whose job was this broad analysis, so we relished a special-interest topic from a Laboratory higher-up, sponsor, or DoD "heavy hitter."

If you are pulling people from a wide variety of groups throughout your organization, you need to convey a sense of important purpose to them. This task will often require the visible support of upper management. I suggest you convince your draftees that they are privileged individuals to be selected for this activity involving key people throughout the organization, working for the director or for some important sponsor. They will get visibility and a chance to enhance their reputation more widely in the organization. Many folks will leap at the opportunity to do something different and challenging for a short while.

All that said, if you achieve a 75 percent batting average on getting full involvement of your team, you are doing very well.

3.1.2 Staffing Your Study Team

Obviously, you look for a cadre of the kind of broad thinkers described in the preceding chapter. You should select at least one promising but inexperienced analyst to begin the "growing" process. You mix in subject-matter experts and hope for an active interaction between these folks.

I have always favored small groups of fewer than 10 analysts and specialists. Small size promotes easy communication and coherence across the whole effort. Your team's task is very much a forward-edge exploration of the problem space, and large numbers of workers are not (yet) needed. You can add in selected in-house experts for specific inputs, but these specialists don't have to be committed to the study. I have found strong administrative persons can be an enormous help to you; they handle the logistics of the meetings and make sure the coffee gets there and is hot! You can then concentrate on the intellectual drill and interaction with your team.

With regard to study duration, shorter is better. With an internal study drill, you have a good chance of capturing your folks close to full time, so it's been my experience that a duration of three to eight weeks would be a reasonable balance.

3.1.3 The One-Week Study

Surprisingly, a lot can be done in a one-week burst of study energy. The problem has to be "right," and your small team has to be equipped with the right backgrounds at the start.

I consider such a study to be a dash through the problem space to scope it out and assess if there are major roadblocks to a more deliberate follow-on pass. In some sense, the one-weeker is a precursor effort to determine if one wants to take on a bigger study enterprise.

A lot of debate and "what if" gets left by the wayside as you speed through the problem. Everyone wants to at least touch the major issues so that you have something coherent to say on Friday!

I have done a few of these one-week studies in my time; one of them is briefly discussed in Chapter 4.

3.1.4 Computer Models and Simulations

Folks today are accustomed to having sophisticated computer simulations and tools available to help with the analysis. In these very early explorations of some new topic, such tools often will not exist, but you can go a long way with intuition, technical and logical reasoning, and pencil and paper calculations where needed. You can do parametric analyses around phenomenological uncertainties. You can make broad and favorable assumptions about defense capability and see if things still look grim for the defense.

In my view, there is no excuse to not proceed. We are the advanced exploration team, and we don't turn back because we lack convenience tools.

War Story

In the mid 1960s, a study I was involved in needed some estimate of the impact of atmospheric nuclear weapon bursts on radar—the so-called blackout problem. I ground through six weeks of radar nuclear blackout calculations, which I essentially did by hand with the aid of a "one-armed bandit" mechanical calculator. The computer simulation guys at the Laboratory told me I was nuts; they would do all these calculations in 30 seconds of run time on the IBM 7094 once they got their program written. They were right about the 30 seconds, but I beat them by two years!

Another advantage of pencil and paper (or chalk) analysis is that it is "scrutable" versus the inscrutability of large computer codes. It encourages focused thinking and team dialog, which are important attributes of a solid analysis. We will have lots more to say about computer codes.

3.1.5 Final Reports

The worst failing in this systems analysis business is to conduct a thorough, thoughtful, and comprehensive analysis and drop the ball on a written final report. A set of final briefing charts is not a final report. In my view, it is not even close!

These analyses we advocate are often the first coherent pass through the problem space. To do that exploration and not leave a quality "map" behind for other investigators is a great disservice, an unforgivable one in my view. Briefing charts don't cut it. Maybe Sherlock Holmes can figure out how you reached many of your conclusions, but the average person following you several years downstream will not be able to do it. Your charts often are not archived, and even if they should be, they will be in the wrong digital media in a few years, and no one will go through the trouble to transcribe them!

There is a tendency to produce the lazy man's final report: a set of briefing charts with words underneath (the so-called annotated charts). I do not think this is a viable substitute. It leaves out too much of the reasoning involved in working your way to your final results. It is better than nothing but, in my view, not much better.

It is all about the archival process of science wherein investigators, years later, can understand your results and how you got them and therefore can assimilate your findings into their own investigations. The medium for this archiving is a written report, and with today's advanced word-processing and printing technology, there is little excuse for not producing a solid report. Your sponsor may only require a final briefing and a copy of the charts, but you owe the community a written report.

3.1.5.1 Scenario

You have finished your internal study, briefed it to top management and the sponsor. Everybody is happy, but your study team quickly disappears back to their day jobs. They have accumulated a severe backlog of work during your study, so the chances of squeezing timely final report work out of them is a definite long-shot.

Well, Mr./Ms. Chairperson, guess who writes the final report? You do! You write it, get it reviewed by your team, and get it printed. I suggest you get on it quickly and work it intensely before other tasks and responsibilities catch up to you. (In the next part of this chapter, I say more about this final report thought process and report production.)

So, along with scoutmaster, ombudsman, and intellectual questioner, you are also the chief writer. Good luck!

Longer War Story

In the early 1990s, the Army had developed a detailed simulation of an enemy concealing his tactical ballistic missiles by movement and hiding. The Army was interested in developing attack capabilities against these missiles, but they were quickly stymied by an inability to model the process by which airborne radars would find the mobile missile launchers. We were asked to help, and I assembled a dream team of in-house expertise. This study was a detailed analysis of automatic target recognition by advanced airborne radars. The Army scenarios were very specific and detailed, complete with well-laid-out threat scenarios with terrain maps and sample missile hide sites. We tried various airborne surveillance strategies and modeled various levels of radar capability. We gave the enemy varying levels of countermeasures. We used several models of an Army ground-launched tactical ballistic missile as our attack weapon.

We produced numerical box scores of enemy missiles killed, false alarms attacked, and enemy missiles successfully launched. Our final briefing was a big hit. I gave it at least a dozen times. Most notable was the set of numerical results that could serve as an R&D guide to developers of automatic target recognition techniques, airborne radar techniques, and countermeasure insights.

But we never produced a written final report! My team disappeared quickly, but I thought I would be able to pull them back together for the writing chore but failed after trying for several months. I still have the briefing charts in my file under "Attack Ops Study." From my viewpoint, this was one of the best analyses ever of the difficult problem of finding mobile missiles, and I as chairman failed to capture the insights for the wider future community. Lesson learned!

3.2 The More Formal Outside Analyses

I have been involved in this type of systems analysis for the past 30 years, so my advice is largely oriented to this type of exercise. My contributing authors tend to focus more on the internal study protocols. I got involved in my first outside study in the early 1960s with participation in a large national study of ballistic missile defense called Project Intercept X. The study was populated with technical experts from a wide range of disciplines, but I didn't think they were paying enough attention to the very broad question of whether the defense radars could develop their information in time for the defense to launch its interceptors early enough to achieve a substantial coverage footprint. So a colleague and I began to focus our work there while the experts argued details of reentry discrimination and detection of warheads. As the study concluded, we found a lot of interest in our results because in some sense they represented the "bottom line." It was a good start for me, and now as I take on some study leadership role, I start asking about the bottom line on day one.

My experience with outside studies over the last 25 years has been dominated by my work on the Defense Science Board, which is a collection of some 40 scientific and engineering leaders from industry and the national laboratories, plus a small cadre of retired senior military officers. This group advises the Secretary of Defense through his undersecretary for acquisition, technology, and logistics. The DSB takes on some 15 to 20 task forces per year to answer particular questions raised by the secretary and other senior leadership. The questions have varying amounts of scientific and technical content, and I have pretty much focused my efforts on task forces with a significant technical content. (I will say more about the increasingly nontechnical study tasks later.) Each task force has a chairperson or, more recently, a set of co-chairs. I have participated in some 26 task forces since about 1985 and have chaired or co-chaired 12 of them. My advice here is based largely on that experience.

3.2.1 The Terms of Reference

The Terms of Reference (TOR) is the one- or two-page document, issued generally by the undersecretary, that establishes the task force and presents the issue to be studied along with a brief background statement on the origin of the issue. It generally provides some guidance on the particular questions to be addressed, along with a suggested schedule and an appointment of the task force leadership and supporting offices within the DoD.

Your first job is with this TOR; you need to get the question right! This TOR gets generated by the DoD staff, and it tends to be somewhat a compromise document among the various staff that generate it. If you are to be in a leadership role on a task force, I advise you pay a lot of attention to the TOR generation process and get involved in its formulation. You need to make sure that the charge to your group is realistic and limited. I have found that at times the DoD staff have a tendency to treat the TOR as a Christmas tree on which they hang their favorite “ornaments” (that is, questions) for you to solve. You need to trim this process back to a set of realistic questions that your group can take on.

Early action on your part is important here; after all, once the TOR is finalized, you are somewhat stuck with it. If you decide later to ignore some of the unrealistic parts of the TOR, you may spend a lot of time explaining why, and you may leave your sponsor with the sense of an incomplete job. The time to achieve a realistic charge to your group is at the start!

3.2.2 Duration of the Task Force

These studies are almost always commuter-style activities with your task force members traveling mostly to Washington and occasionally to a contractor's site or a military facility. Generally, the members’ schedules can only afford a few days a month, so a task force will find itself spread out over months. My experience suggests that a duration of about nine months works well, with an interim briefing to the leadership at about the three- or four-month point (sometimes called a heading check with the leadership). A schedule of two contact days a month with member work in between seems to work well. All this scheduling depends on the nature of the challenge to the task force. I have been on task forces that ran for years as supporting and advising functions to major high-technology programs.

3.2.3 Task Force Membership

The challenge here is often to keep the number of members constrained. A task force on a high-interest topic will have many volunteers and many suggestions by DoD folks as to how knowledgeable and helpful a certain individual would be. I try to view this help with a degree of skepticism; some of these suggested members can have interests to push or axes to grind.

In my experience, the DSB membership was a valuable resource; this cadre was well known to me, they had a vast knowledge of the DoD operations, and they knew the ropes in terms of digging into complex DoD problems. Further, we could reach outside of that membership to tap national expertise as needed; thus, I often got the chance to interact with some of the best minds in the nation.

I favor a task force of 10 members or less. In addition to the members, you will have government advisors and subject-matter government experts plus a few supporting cast members. I would assemble the task force carefully to take on the task at hand with a minimum number of players.

During some intense studies, I found a "utility infielder" to be very helpful. This was a technical person with good organizational skills who assisted me with a lot of details, such as chasing down data, arranging briefs, and back-checking conflicting information we had heard. I was able to focus on the broad conduct of the study. If you are on an intense tight schedule, consider such a person for your team.

3.2.4 Conducting the Study

There are a wide variety of insights and hints I can offer here. Some may work for you as you take on a leadership role in an external study.

3.2.4.1 Panels

Folks have a tendency to set up panels as they organize a study, particularly studies with a large number of participants. You have a threat panel, a sensors panel, a cost panel, a processing panel, and a system panel. Each panel has a chair, and each panel tends to meet separately and develop some input to a final process during which all meet together and hammer out big insights!

I have been on a lot of these "panelized" studies and have serious questions as to the value of this structure. The panel inputs come too late to be of much value to the system panel that has some responsibility to pull together an overall view of the problem. If you are on a panel, you tend to be isolated from the other panels and miss out on their briefings and internal commentary.

I favor fewer people on the study. Individuals can be tasked to bring specific briefings to the whole group, and individuals can be asked to take overall responsibility for, say, threat inputs and making sure the whole group gets the appropriate threat information.

Panelizing tends to divide your group at the start and serves to reduce interaction across the full ensemble of talent you have assembled. This divisiveness is too high a price to pay for a structure that doesn't quite work anyway!

A variation on the panelized study now being used more often by the DSB seems to me to be a good compromise. Panels are formed as focus groups. Study members meet all together most of the time. The focus groups take a higher interest in their particular topic, and they have the responsibility to explore it more deeply if needed and to make sure important facets of their topic are adequately presented to the whole group. This approach avoids much of the separation of the overall thought process that I saw in earlier panelized studies.

3.2.4.2 The Waterfall of Charts

In many high-interest studies, you may get subjected to a seemingly endless set of briefings—a waterfall of briefing charts! Your group can sit there day after day under this barrage of information, some of it useful, some of it not particularly relevant. At the close of each day, I would recommend a caucus of only the study membership to ask, "What does all this mean, and are we getting any closer to the needed insight on the problem?" George Heilmeier's question is appropriate here: "If I had to tell the undersecretary our conclusions today, what would I say?"

Your group caucus will often point to the need for more private group time and less briefings. This input will empower you, the chair, to cut back on the waterfall of charts. It is easy to forget that the most precious commodity you have is this informal group interaction, and it is just too easy to succumb to the waterfall in the belief that some great insight to the study will land on your head!

3.2.4.3 Finding Your Way: The Great Chart

As the waterfall of briefing charts falls upon you, you need to find your way through the mess. Your group's job is one of digestion, sorting, filtering, questioning, and finally synthesizing a broad summary for your sponsor. This summary may manifest itself in a set of words, but often it is a key figure, chart, or table that captures the essence of your story.

Some of the best analysts I have worked with have a magic touch with charts. They seem to know exactly what the form of the key chart needs to be well in advance of being able to construct the chart. My colleague Vic Reis (former DARPA director and former Director of Defense Research and Engineering) is a master at this. He would draw a set of axes and say, "The person who can draw the parametric curves on these axes truly understands the problem." At that moment, he might not have a clue how to draw the curves, but that task became his focus and he solicited inputs from all as he proceeded with his study. Sometimes all he could get were approximate curves for his charts, but he could point to important inflection points and say, "That is where your system will work best."

My greatly admired colleague Dennis Murray, to whom this book is dedicated, was a master at this also. He seemed to think in terms of these "magic" charts, and he did his charts in India ink (a mark of great self-confidence) on thin vellum graph paper. He had a modest-sized briefcase with literally hundreds of these great charts in it, covering all kinds of topics. If a question came up on the engagement range of a particular U.S. fighter aircraft against a particular Soviet fighter, Dennis would pull out a chart from his briefcase and the whole story was there on one chart with maybe five parametric curves. He seemed to pick the parameters and the scales on the axes just right!

It seems to me that engineers and scientists express themselves best with great charts or great tables, so you should be alert for the opportunity to do so in your study.

3.2.4.4 Outsiders, Leaks, and Owls

A high-interest study topic will create demand for people involved in the general topic, but outside the study membership, to attend your deliberations. This demand will make it difficult to maintain a degree of privacy in your study actions. The standard protocol is that your deliberations are private and the first persons to hear your conclusions are the board leadership followed by the study sponsor (the person who asked for your help in the first place), and then, once approved, to the relevant audiences in the DoD.

Control of attendance is therefore necessary, and your study's organization process needs to address this issue up front. One useful protocol is to have relevant outsiders (generally government types) attend while you are receiving briefings since they may be able to clarify confusing situations and point to additional sources of information on your topic. However, when you have your end-of-the-day caucus or other group discussions, you limit attendance to study members (and any needed outsider is included by a specific invitation).

What is an "owl"? This is an individual who was deemed an important outsider attendee at your study by some DoD personage. The owl sits there day by day, never utters a hoot, and never contributes. You will meet some in your days as a study chair.

3.2.4.5 Democratic Operations

Some bright thinkers talk a lot; others are content to not say much. You will get a mix in your study group, and you need to ferret out all opinions with some strategy. My approach is often to ask the quieter ones on a break, "Do you buy all this stuff? Do you agree with our course of action here?" Half the time, they will not be in agreement, and you as the chairperson can put their opinion into play.

A useful approach is the aforementioned informal caucus of the members around the table, allowing each member a say on the study situation at the moment. No matter what you do, the louder folks will tend to dominate, but as chair, you have to do your part for the quiet members.

You may hear some unusual or seemingly weird opinions voiced from time to time by your study members. I would not worry about this too much; the study process will deal with these odd opinions in the course of time. The same applies to strong disagreements among members. You have this forum of the study group and that forum can find its way to reconcile a wide variety of opinions. The process of building the final briefing (to be discussed), which you lead as chairperson, is the final arbiter of disagreements and the way they are handled or captured in the final output.

3.2.4.6 Showstoppers

A few events can slow you down or possibly bring you to a (temporary) halt. You need to navigate around these obstacles.

In a DoD study, special security classification issues can get in your way. The DoD folks may be unwilling to clear your whole panel into some area. This problem should have been handled at the inception of the study by the study sponsor's staff, but sometimes your study course is unpredictable and you bump into the problem along the way. Often, a subset of the panel is given the information, and if that insight becomes substantially important to your final findings, a special classified annex to your final report is prepared.

You can get hung up by some complex phenomenological issue for which there seems to be not enough data or insight to make a solid judgment. You can often work around this showstopper by a parametric process that points to the decisive role of this phenomenon and encourage follow-on efforts to resolve the scientific uncertainty.

Another potential showstopper is the sponsor's request that your study provide cost estimates of some proposed big system under consideration. In my view, the costing of big DoD systems is a decidedly screwy process, and engineers and scientists need to be careful treading in this area! The current process allows no margin for technical uncertainty or surprises, and everyone signs up to a "perfectly executed" program and schedule, which rarely happen. The result: the new airplane costs three times the original estimate, and we are talking big dollar overruns here! So my advice is to avoid the detailed cost drills; get them out of the Terms of Reference. But if the sponsor insists, rely on a team of government professional cost folks and let them use their historical databases on big system costs (and overruns) to answer that mail.

An exception here is the common likelihood that your study will recommend some new-focus, scientific-type investigation. You do owe your sponsor some idea of the level of effort envisioned to crack the problem. In this situation, engineers and scientists are the right cost estimators, and I recall often recommendation words such as "a $100 million per year effort for at least five years."

Whatever the showstopper is, these broad defense system studies we advocate have the ability to navigate around it and provide a useful set of findings and a useful roadmap to the sponsor.

3.2.5 Studies That Involve Nontechnical Analyses and Solutions

I need to digress here to comment on the applicability of the advice we offer herein. There is a proclivity today for various DoD science advisory boards to take on studies of problems that decidedly do not have an underlying core of science and technology. The scientists and engineers of these advisory boards are appropriately responding to the needs and questions of high-ranking DoD officials, but as such, they are functioning, in my view, much like a management consulting firm in commercial industry. Sample questions might be, "How do we achieve more seamless joint operations among our services?" or "How can we increase the mobility of our combat units?" One can find some technical issues in there, but the question cannot be substantially attacked with the tools of engineering, physics, chemistry, or mathematics. Therefore, I argue that these classes of studies are largely outside the scope of any advice we have to offer in this book. We may have some good advice on running a study or finding broad thinkers for your study, but one should be careful to test the applicability of our insights to these more diffuse type questions.

The Defense Science Board is well into taking on these types of questions, and this situation occasioned a dialog on these studies with one board member who has extensive background in DoD management issues, Dr. Robert Hermann, former Assistant Secretary of the Air Force and a longtime DSB member. Excerpts from Bob Hermann's comments on my three chapters follow:

> I did find that, had I written these chapters from my experience, there would be paragraphs on topics that you did not cover. I believe the root cause for these differences in emphasis derives from the fact that I have rarely been involved in a study where the issue to be dealt with called for, for the most part, dealing with deterministic analyses using physical principles. One could decide that my type of study is a different animal and does not qualify as defense systems analysis, but it does constitute a good measure of what the DSB does.

I have chaired studies on Assured Military Use of Space, Dealing with Transnational Threats, Application of Red Teams, Enabling Joint Force Capabilities, and the Military Utility of Broad Area Surveillance, among many. I have also been a participant in many others that often involve more managerial, subjective, and operational judgments not resolved with deterministic logic. I was prompted by your words to think how this might change things. Some thoughts follow.

Consensus is an essential objective. This contrasts with the objective of discovering or developing the deterministic truth of an engineering analysis. It is not as though there is no engineering analysis, but it is generally embedded in issues that are not amenable to deterministic analysis. Therefore, an important characteristic of the result is that a particular set of disparate but respected folks agree on this issue. It is the credibility of the judgment of the group rather than the hard engineering analysis that is the primary prize. This credibility is enhanced if the diversity of the group includes respected members from a wide variety of stakeholders and tribes. ("We need a good Air Force operator for this area.") This diversity goal affects the nature and number of committee members. The personal qualities of the members are important, but so are their heritage and representation value. To cover the constituencies involved well enough may require substantially more than 10 people.

Given a somewhat larger group of disparate peers assembled to capture their collective judgment rather than collectively discover some engineering truths, the dynamics of the group are, in my opinion, somewhat different. I will venture some descriptions of what I mean.

- About marking territory. Inevitably, each legitimately proud participant will, consciously or unconsciously, make a speech or two that will signal to the rest of the group the intellectual territory he or she thinks is his/hers and the intensity with which that territory will be defended. I have found that it is efficient to force that signaling phase early in the exercise by pressing for each member to speak on their going-in thoughts before the scene is corrupted by any data.
- The Committee as an organism. By working out an early collective awareness of who is going to play hard in which territories, the members can better choose where to interject, with what language to yield where needed and press where appropriate, and implicitly pay respect to affected members. I have found that this approach can enable a somewhat self-governing collaboration that has some elements of an organism that is not the creature of the chair or any

other member. This collaboration can be valuable in the end because each member not only becomes an owner of the final product but has arranged the final product such that he or she can represent it to his or her own constituency with comfort.

- Political doability. One of the most frustrating aspects of most studies that I have participated in has to do with each member's conscious or unconscious bias about what things are doable and what things are not. Since no study group is comfortable making recommendations that are in fact undoable, this factor infects almost every aspect of the study. Comments such as, "There is no precedence for that," "DoD regulation does not permit that," "The Air Force would never accept that," "Senator Levin does not want that to happen" and many others occur on a regular basis. In most cases, there is an element of truth in each of the concerns offered, but it has been my experience that trying to address all of a group's individual estimated constraints would render a solution impossible. This topic is worth separate discussion early so that subsequent deliberations are not unnecessarily corrupted.

Bob Hermann's insights above point to the substantially different charters these external studies can have. Again, the advice we offer in this book applies to the studies involving predominantly science and engineering issues, but the reader should be aware of the increasing trend to the type of study discussed above.

3.2.6 Preparing the Final Briefing

In my protocol for these outside studies, the final briefing and the final report are linked together rather tightly. You may find my process useful for your study work or you may devise a process more to your liking. Whatever you do, have a plan for both drills before you start.

One of your challenges as a study chair is to achieve a relatively democratic consensus on the topic from your study group. The quiet folks need to get their two-cents' worth in despite the presence of some decidedly loud competition. I am one of those loudmouths, so I can make this call from experience.

In the good old pre-digital days, I could stand in front of my study team and use the blackboard to ferret out our principal findings and conclusions. A simple drill was, "If we could tell the undersecretary only one result of our investigation, what would that be?" After one hour or so of argument, we would have a good idea of number one. Then on to number two and three, etc.

Note something here: I control the chalk and the eraser. Nothing goes on or off the board unless I do it. I am the conductor of the debate (and I should be careful to never slip into a dictator role). If I don't like what I am seeing and hearing, I can argue or ask questions of the group. I can stand there with a pained look on my face and say, "I am not going to tell that to the undersecretary!"

This style of forum is a democratic consensus process. And quite frankly, I love the give and take. As chairperson in front of the group, you present a big target and welcome their best shots. Believe it or not, you will learn a lot in this initial drill.

After about three hours, you will be tired and ready to call a halt to this process for the day. You promise to be back the next day (or the next meeting) with a briefing of the "blackboard" results.

Your first return briefing try will likely be a contentious process; but you adjust on the fly and persevere to the end with the promise to be back at the next session with version 2. Somewhere around version 3 or 4 of this process, there will be mostly nit-picking of your words, and finally someone in the group will say, "I think he finally has got it," or "He is close." You now have your draft final brief to take forward, and everyone had a chance in a democratic sense to have a say during the tedious process of putting it together.

Now in this digital era, you just replace the blackboard with a single individual running a laptop with the projector. That individual only types what you say or approve. He or she will get deluged with suggestions for word changes, punctuation, and spelling from the group, but you need to keep control of the content. It was easier in the blackboard days; in this laptop era, everybody thinks they are William Shakespeare!

If you want total chaos, let all the laptops come out and generate various versions of the briefing charts. You are now approaching the fabled "one million monkeys with typewriters" creating a great work of literature!

3.2.7 Giving the Final Briefing

I like the protocol used by the DSB for a final briefing. It is given first to the DSB chair and vice-chair. They will be particularly interested in the study quality and its completeness in answering the Terms of Reference, hence my admonition to pay close attention to the TOR content at the start so that you don't get saddled with unrealistic questions to answer. The final briefing then is given to the undersecretary or assistant secretary who requested it. Hopefully, your efforts have "answered the mail," and the briefing is approved for release to designated components of the DoD. The sponsoring executive may request more work, which may require a reconvening of the task force; in most cases, you get the OK to go forward.

If your study was in a high-interest area, you will get to give the briefing a lot, and I mean a lot! I gave the brief of the first DSB task force on GPS (1996) 68 times over a one-year period.

I have said little about the briefing content and how to create actionable recommendations in a compact and interesting format. My contributing authors talk about "telling a good story." A few words of experience, top executives in the DoD are very busy and will rarely sit for a full hour to hear a brief. Target yourself at 30 minutes including some questions and have the famous "two-minute elevator speech" in your hip pocket if you find your time cut drastically short.

Make sure your recommendations are actionable. At times in the past, we were asked to name the DoD component who should be the action office and also to put a cost tag on the necessary action. I never agreed with this protocol. I think it implied too much insight, power, or influence on our part, and I don't think the bureaucracy appreciated a bunch of outside whiz kids telling them how to run their business in such detail.

If at all possible, try to create a memorable chart that captures the result or major insight of the study. The war story below will illustrate both the elevator speech and the killer chart.

War Story

The DSB GPS Task Force of 1996, co-chaired by Professor Steve Koonin (then of Caltech) and me, was briefed very widely in the Pentagon. It contained a classified color chart which had some 15 U.S. weapon systems diagrammed on a technical chart that captured their GPS jamming resistance. Parts of the chart were colored red, and other parts green with an intermediate band of yellow. A system with good jam resistance would appear in the green, bad in the red, and intermediate in the yellow. Some of my study cohorts argued that I had no right to decide on these color grades, but I persisted because one scan of the chart gave the observer a pretty broad sense of the problem.

I got the call to brief Secretary of Defense Bill Perry. I showed up at the assigned time and was told he was running late. The longer I waited, the later his schedule slipped, and I finally met with him at about 8:00 P.M. I knew Bill Perry pretty well, and he really looked tired. I suggested I come back another time, but he wanted to hear the brief since I had waited some four hours. He asked if there were some short form of the brief, and since I knew how technically quick he was, I said, "One chart." I showed him the color chart with the 15 systems displayed; he absorbed it quickly after asking a few questions about the axes and how I chose the color thresholds. He concluded by saying, "Bill, am I correct that we do not have a nightmare here, but we do have work to do?" It was a perfect summary of our task force. I was out of his office in 10 minutes!

Two days later Perry's military assistant called me and said that Dr. Perry had set me up to brief the Joint Chiefs of Staff and, by the way, be sure to show them the rainbow chart!

3.2.8 The Final Report Preparation

After you have given your final briefing about 10 times, you have a pretty good idea of how it goes over, what are the major listener interests, and which areas of the briefing seem to need more elaboration than you first thought. I suggest that your 30-minute briefing charts and the words that go with them are an excellent rendition of a final report. I suggest you sit down and write the damned thing quickly. After all, you have all the words "burned" in your brain and you know where the rough spots are. Your rendition will be in a well-organized flow compared to the alternative of doling out bits and pieces to your study members and then editing the collection of disparate inputs into some coherent whole.

The above process is democratic because your final briefing that guides your writing was democratically generated. I promise you that none of your study members will object as you, their chairperson, take on this task. Of course, you have your study members review and edit the report once, then you reissue it back to them for a second and, eventually, a final review and edit. Your study members have been tasked to provide supporting appendices on topics within their interest or expertise. The details go in these appendices. I like a 20-page final report with 5 to 10 supporting appendices.

Another alternative to producing the final report is one I avoid. Many of these advisory boards have support contractors who provide logistical support to the various task forces or studies. The contractors are very helpful, but they should not be the ones to write the final report. Those words in the report need to be the study membership's words as edited and agreed to by all. Support teams can be a big help with generating figures and getting things printed, but I feel strongly that the words are the responsibility of the study members, and ultimately of the chair.

The protocol I recommended, in which you, the chair, do a lot of the work, is the fastest way to a final report and actually results in less work for you overall. Reconciling different writing styles and managing the problem of one of your study members giving you a very terse one-page write-up of a chapter and another member giving you a veritable book are complex time-consuming tasks. Try my approach; you will like it.

Be aware that the slowest element of this report production process often will be the government. They need security review, review by some other experts, and then printing. None of this seems to move fast.

Overall, you as chairperson need to monitor and drive this process aggressively; it is part of your responsibility. Under no circumstances can you conduct a sound study involving months of effort from some of the best and finest people in the nation and let the final report fall by the wayside!

4

Examples of Defense Systems Analysis Studies

William P. Delaney

I estimate that in the more than 57 years I have been associated with Lincoln Laboratory, I have done more than 100 broad systems analyses, both internal and external. In this chapter, I concentrate on external studies that illustrate some of the points I have made in the preceding chapters. My co-authors include their own examples. My examples contain short renditions of the problem or situation addressed, some idea of the approach, a synopsis of conclusions, and some assessment of the study's value and impact. With the exception of the first example, I chaired or co-chaired the remaining example analyses.

4.1 Two Studies with Enormous Impact in the Early 1950s: The ADSEC Study by the Air Force Scientific Advisory Board and the Project Charles Study by MIT

These short studies occurred as I was in high school; they ran in tight sequence in 1950. It is hard to find a more impactful set of studies since they led to a project of larger scale than the Manhattan Project of World War II!

For the United States, 1949 was a worrisome year: the Soviets had surprised us by detonating an atomic bomb well in advance of our estimates, and we realized they also had bombers that could reach our country. We were already fully into a cold war with them.

The newly formed U.S. Air Force had set up a Scientific Advisory Board under the famous Caltech aerodynamicist Theodore von Karman. Board member George Valley of MIT was familiar with the poor state of our continental air defenses, and he convinced von Karman to have the board do a short study of the situation. Valley was named chair of the Air Defense Systems Engineering Committee (ADSEC) study, which quickly concluded that a major effort to protect the United States against bombers was needed and that the approach should feature a highly integrated system embracing all of the North American continent. Data would be sent from many radar sensors to a limited number of central processing centers to analyze the data, plan a response, and direct that response. At the heart of these centers would be a large digital computer handling the data. Digital computers were quite immature in that era, and the concept of one operating in real time was certainly adventurous! The study pointed to MIT to become involved in this enterprise, as it had run the famous Radiation Laboratory of World War II, which developed microwave radar for the Allies.

MIT responded with its own short study called Project Charles, which concluded that its Whirlwind computer designed for real-time operations to support aircraft simulators could play a central role in a nationwide defense. The study committee argued for a significant R&D effort that would operate off the MIT campus in a dedicated laboratory, and thus was born the SAGE (Semi-Automatic Ground Environment) system and its host laboratory, MIT Lincoln Laboratory. The Cape Cod test system was established along the New England coast from Long Island to Maine with some 20 radars connected by phone lines to a modified Whirlwind computer in Cambridge, Massachusetts.

By 1963, 22 SAGE centers were deployed around the United States (plus one in Canada), thousands of jet fighter interceptors were assigned to Air Defense Command, hundreds of radars were operating, and the whole entourage was connected to a real-time digital system. A conservative office machine outfit called International Business Machines got a kick-start in large main-frame digital computers. (The IBM 700 computer series is directly descended from the SAGE computers.) In 1958, the MITRE Corporation was formed to support the final development and deployment of SAGE.

These two studies lasted less than six months each, had about 10 folks in each, and reign in my mind as the world champs in terms of worldwide impact. Simply stated, they launched the Information Age.

4.2 A Study that Started a Major Long-Term Program: 1977, the Strategic Penetration Technology Summer Study

The U.S. nuclear deterrence strategy rests on three legs of the TRIAD: land-based missiles, submarine-based missiles, and a fleet of bombers designed to penetrate the massive air defenses of the Soviet Union. In 1977, the Air Force had a new strategic bomber, the B-1, in the works. The Navy was enjoying success in developing the Tomahawk cruise missile, which was capable of long-range flight and accurate navigation.

Dr. William Perry was the Pentagon R&D chief in 1977 during the Carter Administration (he would return years later as the Secretary of Defense). A question arose in the Pentagon as to the use of standoff cruise missiles instead of penetrating bombers in the deterrence role. Bomber penetration through Soviet defenses was well studied, but cruise missiles were "new kids on the block," and there was significant debate among various factions as to their survivability in the heavy air defense environment of the Soviet Union. Opinions varied from "the cruise" being invincible to being a sitting duck. Bill Perry called for a study to assess the situation.

I was asked to co-chair the study along with ex-Pentagon professional Pete Aldridge, who was then serving in a Washington think tank (Pete returned to the DoD later to serve as Under Secretary of Defense and later still as Secretary of the Air Force). Pete definitely knew the Pentagon process well, and I assumed my job was to bring some technical judgment to the survivability issues of a modern bomber and the modern cruise missile.

I like to say we recruited the nation's best and finest to serve this study of several months. From Lincoln Laboratory, I signed up David Briggs (later Director of Lincoln Laboratory) and Vic Reis (later Director of DARPA and the Director of Defense Research and Engineering). I recruited the great defense systems analyst Dennis Murray (this book is dedicated to his memory); John Cunningham, an air defense guru of long standing; Stanley Alterman, a renowned electronic warfare expert and innovator; and Paul Howells, a prolific radar inventor and analyst from Syracuse University via General Electric. We had a dream team of talent.

I was captivated by the question of cruise missile survivability and led that portion of the study. Howells, Cunningham, and Alterman focused on the B-1 question. To my surprise, the nation did not have the ability to confidently analyze the survivability of the low-flying, low-observable cruise missile weapon. We were missing some pretty basic science models of masking by terrain, interference by radar ground clutter, and the intricacies of very-low-angle radar propagation. We did lots of calculations on various Soviet defensive weapons by making broad assumptions about these and other relatively scientific effects.

Our study was received favorably. We pointed to the need to get a much better handle on these scientific effects and many other weapon system questions to reduce the uncertainties in cruise missile survivability. I used the study findings as a platform to approach DARPA with the proposition that they start a cruise missile survivability program, and they agreed to a modest effort at Lincoln Laboratory in 1978.

Lincoln Laboratory's early progress in this effort led to our becoming involved in the then very classified stealthy airplane initiative, and we soon had joint sponsorship from the Air Force. That relationship has prospered to this day for some 35 years, and the Lincoln Laboratory Air Vehicle Survivability Evaluation Program has grown into a national effort of significant scale that has involved over a billion dollars of R&D funding over the 35 years.

I believe the 1977 study was the cornerstone of this very successful program; it articulated an important scientific need as the nation pursued new developments in advanced air vehicles. An alert leadership in the Pentagon and DARPA understood the study results and supported the growth of a long-term survivability assessment program under Air Force leadership.

4.3 Providing an Understanding of the Jamming Vulnerability of GPS: The 1996 Defense Science Board Task Force

The Global Positioning System (GPS) caught the U.S. military somewhat by surprise. They had anticipated using it for long-haul navigation, but the rapid improvements in solid-state electronics indicated that GPS could be used in close-in, "hot battle" situations to guide strike aircraft and even individual weapons. In these situations, enemy jamming of GPS would be a significant concern. However, the whole thought process was a bit disorganized; there was a dearth of solid quantitative analysis, and I felt that the military was ignoring the jamming vulnerability (they had not yet experienced any GPS jamming and were disinclined to listen to a science nerd like myself tell them anything about tactical warfare—a common warfighter attitude). I had participated in a 1994 Air Force Scientific Advisory Board on GPS vulnerabilities that included jamming concerns, but the specific jamming insights may have been lost in a host of other concerns about survivability of GPS satellites, ground systems, and the like. The DSB decided to do a study of the GPS jamming issue. (It is often the case that a topic becomes hot, and a wide variety of government studies get launched to investigate the issue.) I would say GPS jamming was warming up around 1996.

I was asked to co-chair the task force along with Professor Steve Koonin of Caltech, a renowned physicist, a very bright individual, and a broad thinker—a type I knew well and could work with. Among the luminaries joining the task force were good friend and always-thoughtful Bob Everett, retired head of the MITRE Corporation, and Stanley Alterman, an electronic countermeasures expert and innovator and "always heard." It was a lively group.

The task force operated for about six months, and in that time we mapped out a quantitative representation of the GPS jamming situation. We found that GPS jamming calculations were easy to do because pretty much everything was known about the system: the signals, the power levels, the receivers, the frequency bands.

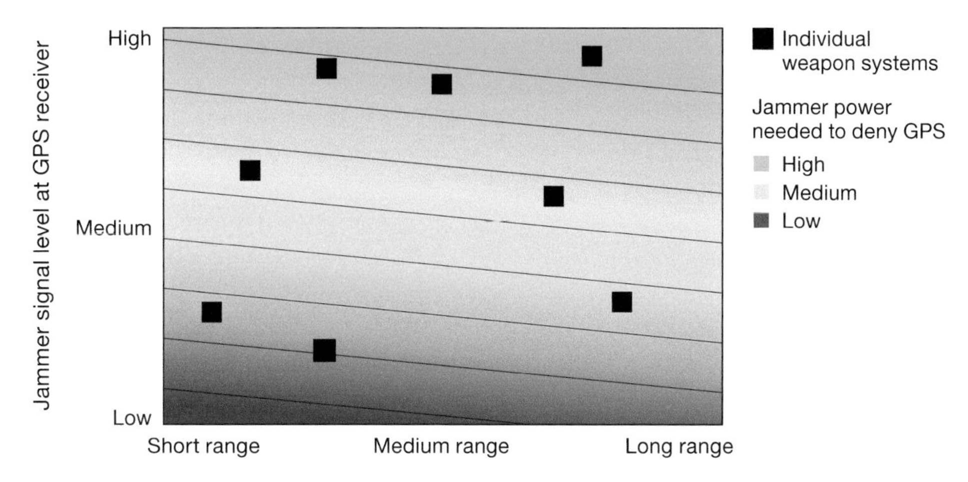

Figure 4.1 An unclassified rendition of the rainbow chart. (Chart courtesy of the Defense Science Board.)

I was fond of saying that you could confidently do the jamming calculation on a single sheet of paper. We focused on simple barrage-noise jamming since its effect was substantial and highly predictable.

Our most prominent result is depicted in an unclassified representation in Figure 4.1. This is the so-called rainbow chart of my war story in Chapter 3. It took some courage to set the thresholds of red, yellow, and green. A system in the red was judged vulnerable because a small-size noise jammer could defeat the navigation mission of the particular platform's GPS receiver. Green meant that a very large jammer was needed, and we presumed that such an observable and lucrative target would be easy to attack physically. The color yellow was the intermediate case in which a jammer of medium size was needed. The chart got a big response in most of our numerous briefings.

Short War Story

A military officer approached me about his weapon system that was on the red-yellow border and said his flag officer wanted their system to be at least in the yellow with a migration path to the green. The chart's quantitative depiction of his system's status made it easy to advise him on a path to follow.

We did identify a wide range of fixes to the GPS jamming problem. We were fond of saying, "The GPS was designed in the early 1970s; let's throw some 1990s technology at it."

The task force's presentation was very popular; I briefed it many times over the ensuing year and still use some of the charts 15 years later. While a task force may do a great job of mapping out a vulnerability in an important system and pointing to fixes, not much may happen quickly. In fact, this DSB task force was the first of four on GPS and jamming that I would chair over the ensuing five years. (A one-week study example is in the next section.) Simply stated, absent some actual military action in which the vulnerability is prominent, it will take a long time to fix even an obvious problem. However, progress in GPS jamming resistance is being made, and I like to think the nagging voice of the DSB over many years is part of the reason.

4.4 A One-Week Study: The 2001 Defense Science Board Task Force on GPS III

By the year 2000, the Air Force had plans for an advanced GPS satellite constellation, called GPS III, with many improvements over the existing GPS. Two major improvements would be a new military signal called M-code and a so-called spot beam on the satellites that could radiate a much stronger GPS signal into a limited geographic area of conflict. Both promised to improve the jamming resistance of GPS, with the spot beam being the big player. Early in 2001, Undersecretary of Defense Pete Aldridge asked for a very quick DSB review of GPS III and the anti-jam value it would provide to U.S. forces. He needed the insights in time for budget decisions, and it seemed like a single week in January 2001 was our only opportunity. I was asked to chair the study and started immediately to think of an appropriate vehicle to display the anti-jam capabilities of GPS III as it was phased in over a 15- to 20-year time frame. (GPS satellites last a long time, so getting some 24 new satellites on station is a many-year process.)

I cheated a bit on the one-week time limit by having some of my best analysts at Lincoln Laboratory, led by Jack Fleischman (a contributing author of this book), do a lot of background calculations on this issue, so I hit the one week in full stride. The baseline approach that evolved was to plot the jamming resistance of a large number of U.S. weapon systems over the many years that GPS III would be deployed.

I fell back on my red, yellow, and green protocol to display how the jamming resistance of the numerous weapon systems improved over the multiple-year window of the GPS III deployment. I put a system in the red if it could not beat a small jammer and put it in the green if it could handle a high-power jammer, and yellow was the intermediate case.

These were not easy calculations since we had to consider many different scenarios for each weapon system. We solved the problem by presenting to the listener four color maps of the multiyear situation. One map, shown in an unclassified rendition in Figure 4.2, was our middle-of-the-road case in which we felt our assumptions favored neither the U.S. weapon system nor the jamming enemy. In another map, the assumptions favored the enemy, and yet another map favored the U.S. weapon systems. A fourth and final map was the baseline case again, but one in which the jamming threat was allowed to grow in intensity over the multiyear interval. Across all four maps, the message was consistent: the M-code would help with jamming, but the spot beam made a dramatic improvement.

I thought it made a great story and I still use it today. As to GPS III progress, the M-code is being deployed, but it seems the advent of the expensive spot beam slips a year each year. Thus, we systems analysts may have to wait a long time to see our recommendations fulfilled, but we pay our way by laying out a clear, quantitative picture for the decision makers.

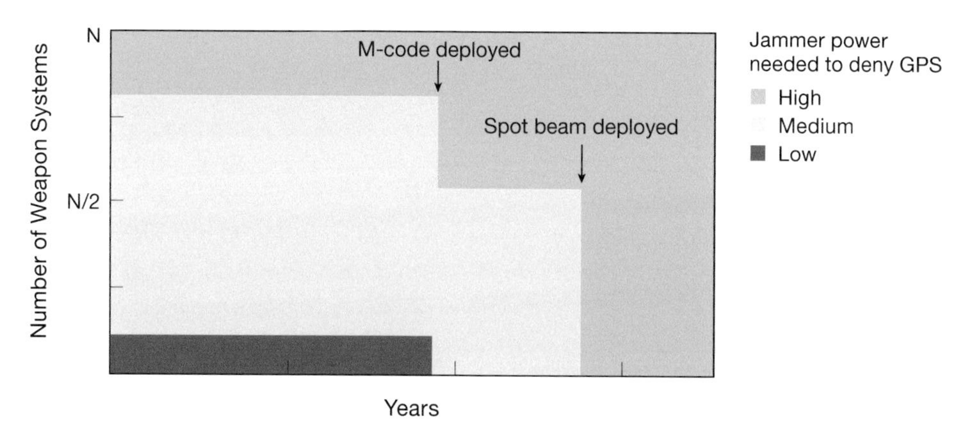

Figure 4.2 Unclassified rendition of the main result of the 2001 DSB one-week task force.(Chart courtesy of the Defense Science Board.)

4.5 Providing a Roadmap for an Evolving Threat: The 1998 Defense Science Board Task Force on Hard and Deeply Buried Targets

An unfriendly Third World country can carry on the development of weapons of mass destruction in sanctuaries from U.S. attack by hiding such developments in tunnels drilled into mountains or in buried facilities protected by many meters of reinforced concrete. These hard and deeply buried target facilities became a particular concern in the era of the first war with Iraq.

I was asked to chair a DSB task force on this issue. I had some familiarity with the topic because of my service on an earlier intelligence community study. We assembled a fine task force consisting of some broad thinkers supported by a collection of experts in buried facilities, weapons effects (conventional and nuclear), and intelligence on these facilities around the world.

Our study lasted about nine months, and quite early we divided the problem into three major tasks:

- How can you find the facilities in the first place? (This is a daunting wide-area search identification process.)
- How do you ascertain the inside features, characteristics, and vulnerabilities of a facility? (This most challenging technical problem involves seeing underground with high resolution.)
- How do you destroy such a facility or otherwise defeat its function? (Hundreds of feet of hard rock can be almost impossible to penetrate!)

We organized our group of specialists and broad thinkers around these tasks and developed a suggested plan of attack for the government. We spent a lot of time with folks from the Defense Threat Reduction Agency (DTRA), which had substantial responsibilities in this area. Figure 4.3 shows our task force and the DTRA support team outside a test facility in Nevada created by the massive tunnel-boring machine in the background. As a kid, I had harbored an ambition to be a civil engineer with a specialty in explosives, so this was a particularly fascinating investigation for me.

The report on this topic is classified, but the Pentagon's principal contact on the problem stated that our final report became their program plan. One of our recommendations was that the DoD further develop the concept of a special 30,000-pound deep penetrating bomb then called Big BLU. (The BLU-109 was our principal 2000-pound penetrating bomb.) The Air Force did not pick up on that recommendation for many

Figure 4.3 Task force and Nevada test site hosts in front of a tunnel boring machine (1998). (Photograph courtesy of the Defense Science Board.)

years, but I am pleased to note the current testing of such a weapon now called the MOP (Massive Ordnance Penetrator). Figure 4.4 shows the MOP in a test release from a B-52 aircraft in 2009. It seems we were on the right track some 11 years earlier!

4.6 A Study with a Particularly Easy-to-Understand Result: The 2003 DSB Task Force on Unexploded Ordnance

Millions of acres of land in the continental United States are unusable because of the possible presence of unexploded ordnance left over from testing as far back as the Civil War, but mostly from World War II. The attempts to remediate this sometimes very valuable land were going slowly in 2002 when I was asked to co-chair a DSB task force on this issue along with Professor Delores Etter of the Naval Academy (later to be Assistant Secretary of the Navy).

This study was about a nine-month investigation, and our small group became familiar with the difficulties of finding and clearing ordnance on a large scale. There

Figure 4.4 Massive Ordnance Penetrator test release from a B-52 aircraft in 2009. (Photograph courtesy of U.S. Air Force.)

was substantial incentive to do the clearing because some of the land was extremely valuable for development. A good example is Fort Ord in California, just north of Monterey on the California coastline—some of the most valuable real estate in the United States.

The problem is that potential explosives are underground, and our ability to confidently identify objects underground is weak. A rock can be confused with a hand grenade, any metallic object is suspect, and military test ranges tend to be cluttered with metallic debris. Numerically, the process is drastically impacted by a horrendous number of false alarms wherein a detected harmless object (a buried soda can) must be treated as a lethal object until it is excavated and identified. It costs $150 to dig such a hole and remove the detected object. At the then-current false-alarm rate applied to the two million acres of affected land in the United States, one would have to dig 200 million holes at a cost of $30 billion. Figure 4.5 shows an example of the extreme number of holes one may have to dig based on a survey by ground-penetrating sensors.

The obvious recommendation was to spend R&D dollars to reduce the false-alarm rate. Task force member (and contributing author) Robert Stein produced a simple

Figure 4.5 A mass of flags marks suspicious underground objects at a test site. (Photograph by Dr. Jeffrey Marqusee courtesy of the Office of the Secretary of Defense.)

table to make the case for more R&D funding on false-alarm reduction. We judged that an enhanced R&D effort, estimated at $100 million a year, could produce a tenfold reduction in false alarms (a reasonable goal for reduction by using higher-resolution sensors and more sensor modalities); the total job would then be $3 billion. A simple saving of $27 billion! It is pretty difficult to come up with a stronger incentive than that!

Frankly, at the conclusion of our study, we did not have high hopes that the government would move in our recommended direction, and they seemingly did not. But in 2011, I heard from one of the researchers in this unexploded ordnance field that they were working the false-alarm problem hard with favorable results. He said our report had made the difference!

Sometimes in this broad analysis business, you do not see direct or prompt response to your recommendations, but eventually they have an effect. Often you are acting as a blocking back for good folks in the government, and you help them get to where they know they should go!

4.7 A Forbidden Study: A 1980s Era Informal Study on Unconventional Single-Silo Missile Defense

My last example is more of a war story than a report on an official study, but it points out that systems analysts can have the opportunity to exercise more vision than your standard government program office.

Active defense of the Minuteman missile silos in the central United States against a disarming Soviet ICBM attack was a hot topic in the 1980s. The very "hard" missile silo was a "nice" target to defend because your defense goal was only to modestly raise the price to kill the silo-based missile, and the defense could be a very close-in, last-ditch style of defense that for once seemed to favor the defender instead of the attacker (to be contrasted with city defense in which only one warhead leaking through had to be considered a drastic loss for the defense).

Various defense architectures were being debated by the Army folks in charge of missile defense. The architectures ranged from a defense unit per missile silo up to a deployment of defense hardware across a field of Minuteman silos protecting, say, 25 to 100 silos. The missile defense analysts in our group had a gut feeling that single-silo defense could be an effective approach. The hardware of very small radars and short-range interceptors was attractive, and the defense goal of killing only one or two incoming warheads before being defeated was a pleasant change from classical missile defense requirements of essentially flawless operation through an extended and intense battle.

The Army decided to focus on multiple-silo defense architectures, and researchers were told to cease work on single-silo approaches. The top management at Lincoln Laboratory made sure we heard the message.

Our small systems group was unhappy with this dogmatic approach to such a challenging defense issue. We harbored an intuitive belief that one could design an unconventional "cheap and dirty" single-silo system that would force the price to kill a silo above one very accurate enemy warhead. Most of the unhappy folks (including myself and contributing author Steve Weiner) played bridge at lunchtime. We conjured up the rationale that we give up bridge for a month and spend that time doing a quick study of unconventional ways to do single-silo defense. Somehow doing the work at lunchtime bridge was judged to free us from the Army edict—don't ask me to explain that logic! A better rationale was scientific curiosity in action.

We chased down five or six ideas—all of them a bit bizarre in concept. "Bomb on a stick" featured a tower out in front of the silo in the approach corridor of Soviet warheads. Our own nuclear warhead was on the tower, to be detonated when the enemy warhead passed nearby. The arrangement was geometrically set up so we could kill the passing warhead but not our silo.

My favorite scheme was "Delaney's rock pit," wherein a large pit in front of the silo (in the threat corridor) was first loaded with tons of high explosives followed by many tons of one-pound rocks. This arrangement formed a crude but huge shotgun device. The explosive was ignited at the right time to fill the sky with high-velocity rocks to impact the incoming warhead and kill it. Lots of physics in there for everybody (velocity of the rocks, survivability of the rocks during "boost," warhead kill by impact, etc.). We also considered at least one short-range classical gun or cannon approach, plus a few others I can no longer remember. All schemes had lots of uncertainties, and none stood out as a winning idea from a defense viewpoint, except that an attacker demands very high confidence and any of these kooky ideas would probably make him double up or triple up his attacking warheads.

Remember my preaching in Chapter 3 that you always write a final report? Well, despite the ban on this class of study, we wrote a single copy final report, put it away in the files, and said nothing to anybody. Fast-forward about a year and the Army decides to relook at single-silo defense! An Army official visits the Laboratory and asks for ideas on single-silo defense. Our assistant director reminds him of the earlier edict and tells him we have nothing to offer. Steve Weiner and I are sitting there, and the assistant director says to us, "We have nothing to offer—isn't that right?" I said, "Well, there was that drill we did instead of lunchtime bridge for a few days." Steve retrieves our report, and we subsequently brief the Army sponsor on our findings.

Single-silo defense never went anywhere, but I have always felt good about out-guessing the leadership. Actually, we were not guessing at all; we were just resisting some less-informed authority's ability to restrict the playing field!

About the Author

William P. Delaney has had a 57-year association with MIT Lincoln Laboratory and is currently the Director's Office Fellow. He is a former Assistant Director of the Laboratory, with experience in missile defense, air defense, radar, GPS, and battlefield surveillance. He spent a three-year tour in the Department of Defense with oversight for R&D on missile, air, and space defense, and a two-year tour on Kwajalein Atoll in the Pacific, leading the team that built the first long-range wideband radar. He established the major program at Lincoln Laboratory that investigates modern air defense problems (referred to in this text as the Air Force Red Team). His education includes RCA Institutes, 1953; Rensselaer Polytechnic Institute (BEE), 1957; and MIT (MSEE), 1959.

He is currently a member of the Defense Science Board and serves on a variety of Defense Advisory Panels. He is a Fellow of the IEEE and a recipient of medals from the Office of the Secretary of Defense, the Air Force, and the Navy. In 2012, he was elected to the National Academy of Engineering for his contributions to radar for national defense.

II

Broad Views of Defense Systems Analysis

5

Some Thoughts about Systems Engineering, Analysis, and Practitioners: An Industry Perspective

Robert Stein

After more than 50 years in systems engineering and analysis, I look back on how I got into it, a few of my most formative experiences, and "giants" who helped shape my thinking, and offer some advice for young systems engineers who hopefully will become the next generation of this increasingly scarce, most-needed, and very rewarding discipline.

5.1 The Beginning: If I Couldn't Do Real Electrical Engineering...

"Maybe you should consider being a systems engineer." I was a young hardware design engineer, and these words were spoken to me in 1961 by my boss, a man who 30 years later would become CEO of the company. I was in a screen room debugging a front-end radar receiver breadboard I had designed and built. The first stage was mistuned a little, and I needed to slightly adjust the reactance of a coil I had fashioned out of a few loops of stiff wire. No matter how hard I had tried, slightly opening and closing the loop spacing hadn't worked because the change was too great. Looking around the screen room for something else to try, I spotted an old cigar butt sitting in an ashtray. The end was well chewed, meaning that there was a lot of moisture in it and its diameter was about the size of my coil. Using it as a tuning slug, I inserted it into my coil and, while staring intently at my oscilloscope, I moved it slightly up and back in the coil until I got the response I was looking for. Unfortunately, as I was totally absorbed in what I was doing, I didn't realize that my boss had walked up behind me, watched what I was doing for a few moments, and made a quick judgment about my talents as a hardware design engineer, prompting his comment. Although he was joking at the time, we both knew that there was truth in the observation, and I later gave up my soldering iron in favor of a slide rule, then a Friden adding machine, then an electronic desk calculator, and then, over the years, with whatever evolution of electronic calculation was available.

5.2 Systems Analysis: My View

After 50 years, I have finally been able to define the word *system*. For the design engineer, it is the scale of the problem one level above whatever he or she is working on at the moment. If the engineer is working on a microwave power amplifier, the antenna module into which it will go is the system. If the work is on the module, the phased array antenna is the system. For the antenna, the radar is the system, for the radar the integrity of air defense is the system, etc. This perspective then leads us to the definition of a system of systems: it is the scale two levels above the engineer's current activity. For a systems engineer, you can subtract one level from the definitions above.

I gradually learned about the progression and breadth of what constitutes a system. But what was in my head from the beginning, and the issue that always held my attention the most, was my fascination with exploring alternate ways of satisfying an objective or requirement, how I might factor in constraints, what constituted a "best" approach, why that was best (the understanding and formulation of this into value structures would come later), and how I could wrap all this into easily understood and explainable arguments. I believe that this fascination with high-level alternatives and trade-offs, and with broad measures of good and bad; the ability to understand the differences between fundamental issues and details; and an inherent dislike for digging out the last few tenths of decibels in favor of establishing gross trends are all common attributes of good systems thinkers and analysts. This is not to say that details are not required; they are, at some level. The best systems analysts seem to know instinctively how much detail is useful, how much is worth pursuing, and how much actually detracts from a clear portrayal of the important findings. And the grounding in reality provided by a few years as a design engineer building real hardware (there was no such thing as software at the time) created a useful perspective on the art of the possible that many of my analytical colleagues lacked.

5.3 Cutting My Teeth on Radar Signal Processing Analysis

My first systems assignment came a few months after the screen-room cigar incident. If we apply my definition of a system above, since I was a receiver designer, a radar was the system. Two new cutting-edge technologies had just made their appearance in advanced radar tradecraft—pulse compression and monopulse angle tracking. Since I had always had a mathematical bent, I was given the task of analytically understanding the benefits, limitations, and potential applications of these two technologies in a variety of airborne radars that we were producing. I was to work with the marketing person who had responsibility for this radar line, so between the two of us, we could examine the market potential, the mission benefits that these techniques could provide, the receptivity of the customer to new capabilities and enhancements, and an assessment of what was technically possible based on the physics of the problem. I was to examine the technical part and my marketing colleague (far senior to me in age, experience, clout, and in every other way) was to provide everything else. It turned out to be my first experience with "technical isn't everything," and, in fact, sometimes not even anything.

A number of articles had been written on these two technologies, and I read them eagerly. I became familiar with maximum-likelihood estimators, the difference between resolution and accuracy, and the reality that you couldn't beat the fundamentals: signal-to-noise ratio, bandwidth (in the case of pulse compression), and the ratio of antenna size and wavelength (in the case of monopulse angle estimation). But perhaps the most important lesson I learned as a budding young systems engineer was that there were those who wanted to sell and those who wanted to understand technical truth, and satisfying both desires was sometimes difficult. It was a lesson that, as the years went by, shaped my ability to use analysis constructively to serve the interests of advancing the competitiveness of a concept while maintaining technical integrity.

Having given up the soldering iron (engineers actually laid out chassis, soldered components on boards, and connected them together with wire) in favor of the slide rule and Friden adding machine, I was now looked upon as a systems engineer. Systems engineering had not yet become as disciplined or as complex as it is now, and the distinction between systems engineering, systems analysis, operations analysis, etc., had not been established. Requirements flow-down was usually done at the back-of-the-envelope level, at least in the case of subsystem definition as in a radar; interface control documents had yet to be invented; functional analysis and partitioning were likewise more seat-of-the-pants than established tradecraft; and it all blended into the general area of systems engineering. Although I'm not sure I ever saw a definition of what constituted systems engineering, to me it was crystal clear that it was neither electrical design nor product design, the other two major engineering disciplines at the time.

As it became apparent to both me and my bosses that I was much more suited to analysis than to hardware design, I found that a vocation in which I could erase my mistakes rather than unsoldering them was well matched to my temperament. The breadth of the analysis in which I became involved steadily increased. I had always been fascinated by a couple of senior systems engineers who seemed to be at the forefront of all new radar pursuits. At the first technical kickoff meeting, they would appear, as if delivering tablets from the mountain, with a listing of the major high-level parameters that would define all subsequent work: search volume and frame time, track update rates, peak power, average power, operating frequency, aperture area and dimensions, instantaneous bandwidth, pulse widths and repetition frequencies, waveforms, etc. Where did these parameters come from? Why were these the right ones? Were there alternatives that they had looked at and rejected? What drove the selection? What if they had chosen a different set? These and a host of other questions filled my head.

As I began to feel more comfortable in this new environment, I started to ask some of these questions. Soon, either to shut me up or because I was asking questions that no one else seemed to care about, I was assigned to work with these two systems engineers. I became their desktop calculator, and they became my mentors. I saw the world of high-level trade-offs for the first time, how they were defined by requirements established by the customer, and most fascinating of all, "bucket curves." The ordinate contained some important factor, such as cost, or an attribute that was really important, such as radar size and weight in the case of truck-mounted radars, and the abscissa contained one of the major parameters of the radar, such as aperture size. Lo and behold, when the relationship between the two was established and plotted, there was a low point in the curve with increases to either side. Just two or three of these curves, along with lots of discussion on corollary issues such as prioritization, were sufficient to define the radar system. Once one arrived at a high-level definition, the rest of the parameters (the so-called derived requirements) just fell out. I was now being exposed to analyses at a level much higher than processing techniques, and I ate it up.

5.4 Moving on to Overall Radar Systems Analysis

After working with my two mentors for a year or so, the company had an organizational change, and a few of my colleagues, my boss, and I moved to a different division. The move did not include my mentors, and so I became the de facto radar systems analyst of record. We were now in the Radar Systems Department, and my boss was the section manager of the Range Instrumentation Radar section. He had come up with a clever idea for a phased array radar (antenna phase-shifter modules were in their infancy) to measure ballistic missile reentry phenomenology. The concept was built around an array with three concentric areas of phase shifters: an inner area with tightly spaced elements, a middle area with more widely spaced elements, and an outer area with still wider element spacing. The idea was that for objects that were very far away, the entire aperture was required to get the needed sensitivity, but because things were far away, the scan angles required were relatively small and thus the element spacing could be large. In contrast, for objects that were fairly close to the radar, sensitivity could be relaxed and only the inner area of elements was required. Because the observations were close to the radar, the angular field of view was much greater, thus the tighter element spacing.

It was a unique concept for the time. We traveled all over the country, briefing and attempting to sell the concept, and I interacted with all of the major advanced ballistic missile defense (BMD) study organizations and potential customers of the day—Advanced Research Projects Agency (ARPA, now DARPA), the Aerospace Corporation, Lincoln Laboratory, Cornell Aeronautical Labs, Riverside Research, Bell Labs, the Space and Missile Systems Organization (SAMSO), the Office of the Director for Defense Research and Engineering (ODDR&E), and a host of others whom I have long since forgotten. In terms of systems analysis, this was a rich time of training and learning. I was now responsible for all of the analyses—waveforms, radar range equation, ambiguity functions, processing, and error analyses—and each presentation we made raised a variety of questions. As I briefed the technical part, I either answered on the spot or promised to return with a more complete answer. I quickly learned that saying I didn't know the answer but would analyze it and follow up was far superior to faking an answer, flunking the follow-up questions, and losing all credibility.

The experience taught me how to do quick-turnaround analyses. My boss's standard rule was that whatever answers we promised had to be available within a week of our promise. Long, drawn-out analyses seeking truth to a fraction of a decibel didn't fit this paradigm of getting a question, providing a partial answer, and getting back in a week with a more complete answer. These two lessons—don't fake answers you don't have, and back-of-the-envelope answers are good enough as long as they treat the critical issues—served me well for the rest of my professional career.

I also learned about how to put together briefings. I discovered that there was an important middle road between including only the most critical, high-level issues without any supporting analysis and providing so much analytical support that the forest got lost because of the trees. I learned that the balance depended critically upon the audience, and each briefing had to be tailored to the specific recipients. Too much detail and many audiences missed the message; too little detail ended up with the appearance of arm waving. Most other good analysts knew instinctively about this balance. In the early 1970s, Steve Weiner—one of the most accomplished analysts I have ever known and a cocontributor to this book—after sitting through a long and detailed presentation to a Naval Research Advisory Committee study filled with rambling backup and so-called data, stated his view that "study viewgraphs had to contain sufficient detail to establish credibility. If the detail related to the subject at hand, so much the better." That tongue-in-cheek observation, as so many others of Steve's, seemed to sum things up very nicely and hung over my office door for the next 30 years.

There was one other major lesson learned: if you are the one who will brief the analyses, make sure that you anticipate as many nasty questions as possible and have the answers to them in your pocket. In effect, I became my own red team and murder board, long before these were institutionalized within the company.

5.5 Weapons Systems Analysis and the Most Analyzed Enterprise of All Time: Missile Defense

However extensive this systems analysis learning experience was, it was still largely constrained to radar systems analysis. This level of analysis broadened significantly over the next decade or so when I became heavily involved in one of the major Defense Department thrusts of the late 1960s to mid-1970s, the active defense of the country's ground-based nuclear retaliatory capability. There were two primary interrelated architectural issues around which years of cost-effectiveness analyses churned: (1) how many Minuteman silos should an individual defense site protect, and (2) to what degree, depending upon the answer to (1), did the defense sites need to make sure that they survived preemptive attack? The requirements were simple (e.g., ensure the survival of 300 out of the 1000 existing silos against an attack of 2400 Soviet warheads), but the analyses were complex.

On one end of the spectrum, small, "inexpensive" defense systems each defended a single Minuteman silo, and defense survivability was of little concern. At the other end of the spectrum, Safeguard, for example, a single site defended many silos, and therefore the offense obtained lots of leverage by killing the defense before attacking the silos. Here, defense survivability was the dominant issue. In between, study after study extolled the virtues of their recommended "N on M" construct, with reams of analyses to prove their point. Various concepts for defense survivability were conceived, e.g., the Bell Labs Virtual Radar Defense (VIRADE), in which the defense was on a train that moved around the Minuteman fields so that offense targeting was difficult; an ARPA concept for putting the defense in empty Minuteman silos and moving them around from time to time; and a concept for fixed buried interceptors, with fire-control radars on transportable trucks. The analyses on survivability, offense tactics, defense tactics, radar sizing, missile sizing, the avoidance of missile fratricide with various intercept planning schema (we were using nuclear-tipped interceptors at the time), engagement logic, and overall cost were an analyst's dream—many dimensions to analyze, all of them interdependent, and some of them nonlinear. Equally exciting, there were always eager, high-level, and very knowledgeable government audiences ready and willing to listen and express their opinions to all briefers and concepts.

It was thus a heady time for young analysts like myself, and we all were exposed to various aspects of the analysis trade. I learned about game-theoretic solutions and the analysis of zero-sum games. Since ensuring the 30% survival of Minuteman silos was about deterrence, I learned about analyzing effectiveness from both a defense conservative standpoint (the defense's surety that no less than 30% of the silos would survive under all of the statistical conditions in which things might go right for the offense and bad for the defense) and from the offense conservative standpoint (the offense's surety that no more than 30% of the silos would survive under all of the statistical conditions in which things might go right for the defense and bad for the offense), and I learned how to choose appropriate strategies in between. There was a three-day ballistic missile defense symposium at the Naval Postgraduate School in Monterey every year, and a good 25% of the papers presented were on an aspect of this strategic analysis and the trade-offs leading up to the presenter's preferred solution.

Because everyone in the business was doing these trade-offs and analyses, the government's Advanced Ballistic Missile Defense Agency (ABMDA) decided that it should hold an industry-wide study to evaluate the cost-effectiveness of various Minuteman III active defense alternatives, pick a preferred construct, establish system- and subsystem-level derived requirements, and form a baseline for future industry competition. The effort was called the Minuteman Defense Study (MDS), and all of the BMD analysts from industry, study houses, federally funded research and development centers (FFRDC), and ABMDA participated. The study group met for two or three days once a month, with assigned homework in between, and the venue was changed for each meeting to spread the burden of travel among participants. The study went on for about a year, but never quite produced a clear baseline or a set of requirements sufficient to establish a framework for follow-on competitive efforts. As one observer later described these meetings, "Each participant came into the meeting with a bag of uniquely colored marbles. As each person made a point, he or she put a colored marble in the middle of the table. By the end of the day, the table had a large pile of different-colored marbles all mixed up in the middle. The following day this continued and the pile of marbles grew larger. At the conclusion of the meeting, each participant leaned over the table, picked out his or her own particular marbles, put them back in his or her bag, and went home. At the next month's meeting, the process repeated itself."

One of the reasons for the lack of a definitive study conclusion was the fact that all of the industry participants had a competitive position they were protecting. Although all of the sessions were filled with very rich give-and-take in terms of the systems analyses that were being presented and shared, very little of the discussion had any lasting effect in moving the participants off of their initial positions aimed at protecting what they viewed as their competitive advantage. Thus, although there was a seeming atmosphere of all working together to seek "truth" based on well-founded analysis, that truth was still in the eyes of the beholder and based to a large extent on maintaining a competitive position. A year later, the study name was changed to MDS-1 because a second study, MDS-2, was initiated that was supposed to home in on the subsystem definition and requirements (radars; missiles; launchers; command, control, and communications (C^3); battle management software; system logic; and algorithms; etc.), but its success was as marginal as that of its predecessor.

This experience was rich with good analyses and systems engineering. It taught me a great deal about doing analysis in a group study environment and interacting with other analysts who were technically more proficient than I was and much more experienced in this sort of group's social dynamics. However, the two studies did little to define a common best approach for a cost-effective defense of our nuclear deterrent. The government's solution to this impasse was to embark on a competitive concept definition for this mission, known as hard site defense (HSD). Three primes were each awarded a yearlong study to define their own preferred approach. These studies entailed defining a concept; system and subsystem level specs; a hardware and software integration, test, and evaluation plan; the identification of the equivalent of today's critical technology elements; a risk identification and mitigation plan; all of the standard management plans; and a proposal for full-scale development. Based on a complicated evaluation process, the competition was to result in a winner-take-all for development and production of the system.

Extensive teaming was involved on all three primary teams. One of our team members was General Research Corporation (GRC), a small analysis company in Santa Barbara, California. GRC had a wealth of very talented and experienced analysts and was to form the core of our systems analysis effort to define the overall system architecture, help conduct the system-level trades, and perform the initial battle management, system logic, and algorithm development. As the person leading our team's analysis efforts, I went out to California for over six months to oversee GRC's work and coordinate the effort with everything else that was going on back on the East Coast.

5.6 Systems Analysis Mentoring from the Best

The HSD system study was another exciting learning experience for me. I was immersed in an environment in which analytic integrity appeared to be the driving force and was surrounded by systems analysts and engineers who collectively had expertise in virtually every discipline associated with defense matters. The analyses that were being conducted covered a host of system engineering issues—radar sizing, missile sizing, computer sizing, discrimination, force-on-force analysis, game theory, nuclear effects, battle management, lethality—and each day fascinating discussions would take place on the impact of one or another of these domain analyses and some other seemingly unrelated issue. This, of course, is the essence of system engineering and analysis—understanding and addressing the impact on the whole of issues related to each of the constituent parts or phenomenologies.

But beyond all the engineering and analysis, the thing that both impressed and taught me the most and left the greatest impact on my future as a systems engineer was the presence of Ben Alexander. Ben was the president of GRC at the time, had been Director of Research in the DDR&E at the Pentagon a few years before, and was highly respected, if not revered, throughout the defense community for his intellectual capabilities and his ability to help engineers understand what they were doing and how to do it better. He was by nature a gentle individual, but the power of his intellect, grasp of technical issues across many different domains, and ability to see the whole through myriad pieces was unparalleled. He thrived on interacting with young, inquisitive engineers and, in his gentle supportive way, pushed them to question everything, especially themselves.

My schedule for eight months was to fly to Boston on Friday evening, attend a staff meeting on Monday morning to deal with and consolidate issues raised the previous week, and fly back Monday evening to Santa Barbara, where I would work for the rest of the week. Understanding that the Monday staff meeting was the critical time to raise important new technical issues and be able to describe the potential impact on the overall system, Ben would hold his own review first thing Friday morning, and we all would spend the entire day discussing the high-level significance of whatever new technical findings had been raised earlier that week. Ben would lean back and gently question, challenge, comment, discuss, and often raise opposing interpretations, but always as a way of ensuring that all alternatives were considered rather than dictating a personal view. It was an ongoing lesson for all of us on a variety of systems engineering leadership qualities—technical breadth and depth, the ability to always keep one's eye on the forest when dealing with details in the trees, the art of gentle persuasion, and, most of all, maintaining technical integrity. I would leave for Boston on Friday evening armed with two or three new issues that were important to our HSD system concept definition, why these issues were important, what their impact on the overall system might be, and what actions we should take to either explore further or incorporate these new findings. It made me sometimes the star and sometimes the villain of the Monday morning staff meeting, but the reality was that Ben Alexander, not me, was the true star.

In the end, we did not win the Engineering Development phase. Our loss was attributed to something that is almost unfathomable today. It was the early 1970s, and one of the most critical issues in HSD was the development of a computer that had a processing throughput of about 20 million instructions per second (MIPS) and the ability to write efficient software for such a powerful machine. By today's standards, this situation seems incredible. The MIPS measure is not useful today because of the wide variation in instruction sets, the impact of things other than the processor on throughput, and a host of other issues; however, even in gross comparison, 20 MIPS is today far below any meaningful threshold of computer performance. Current PCs that can be purchased for a few hundred dollars operate at instruction rates in the tens of thousands of equivalent MIPS. Yet 20 MIPS was a major technical challenge for a mainframe digital machine in the early 1970s.

We had made the strategic decision to team with IBM Federal Scientific Division for software, but to compete the purchase of the mainframe between IBM, CDC, and Univac, each of which had a developmental computer around 20 MIPS. During the evaluation, we later learned, the issue was raised as to whether or not we could do a fair evaluation of the hardware alternatives with IBM on our team because our selection would be influenced by the risk in having IBM program somebody else's mainframe. Thus, we projected a strong possibility that we would either shy away from a fair and open computer selection or end up with a very risky software development. As impossible as this might seem as a critical issue today, it was one of the dominant factors in the evaluation.

Although we lost the HSD competition, the president of our company decided that in the conduct of the HSD competition, we had put together a cross-company systems engineering team with multidisciplinary expertise that was unprecedented in the history of the company. To avoid losing this, he established the core of the HSD systems engineering team as a high-level group to lead broad system technical efforts for future company pursuits. I was lucky enough to be one of the half-dozen engineers in that group and equally lucky to be working for the man who was selected to lead that group, Mike Fossier, a company vice president who was known throughout the defense industry as one of the brightest, broadest, most technically capable leaders in the field. As I learned throughout the years that followed, his reputation was fully deserved, perhaps understated, and he was one of only a handful of people I came to know during my 50 years in the business whom I would describe as in a class of his own. These people include Mike; Ben Alexander, whom I have already mentioned; Gene Fubini, whom I will talk about later; and Johnny Foster and Dennis Murray, both of whom I have had the pleasure and benefit of working with on various studies over the past 20 years. All except Mike and Dennis have at one time or another held leadership positions in ODDR&E, and all, without exception, have had a major beneficial influence on numerous technical issues over the years.

5.7 Multimission, Multidimensional Systems Analysis and the Notion of Flavors

A year later, as part of this newly formed high-level systems engineering group, I was involved in a major systems engineering experience with the Navy's Shipboard Intermediate Range Combat System (SIRCS) competition. The DoD was about to implement a new acquisition initiative[1] aimed at increasing innovation, examination of alternatives, and industry competition in developing solutions to major agency needs. Created by the Office of Management and Budget (OMB) and applicable to all government agencies, the initiative was described in OMB Circular A-109, dated April 1976, and had at its root three major tenets:

1. The agency had to specify the problem it was trying to solve, not a specific solution or implementation.

2. Competition had to be maintained as long as economically feasible, at a minimum through concept definition and advanced development. The government was to take great care in its dealings with each contractor to avoid prejudicing that competition in any way.

3. Each prime had to thoroughly explore alternative solutions, identify quantitative pros and cons of each one, and perform an overall evaluation of the pros and cons leading up to that prime's selection of a recommended baseline.

---

1 "New" acquisition initiatives have been the norm in the U.S. government and, in particular, the DoD every few years for at least the last 40 years or so. Each one has been an attempt to "fix" one or more specific problems evidenced at the time by any or all of the following: excess cost, excess risk, lack of competition, lack of innovation, lack of supportability, lack of meaningful alternatives, failure to meet user needs and requirements, excess development time, etc. In the opinion of this author, none of these initiatives has resulted in meaningful reform, but just has had the tendency to add more bureaucracy, steps, and process to an already overly burdensome system.

The SIRCS competition was to be the first major DoD acquisition program under the A-109 process, and the Navy was eager to show that they were up to the challenge. SIRCS was supposed to resolve the lack of a fully integrated, multimission, intermediate-range combat system on destroyers. These ships had three major surface combatant missions to perform—air defense, shore bombardment, and surface-to-surface combat. Up to that time, the three were treated as independent missions on destroyers, and all kinds of limitations existed on the ability of the ships to perform any of the other two missions when they were engaged in combat on the other one—there was no integrated combat control, radar resources were not managed as multimission, the radars themselves lacked the capabilities to service all three domains, the use of one type of weapon would constrain the firing of another, and on and on. SIRCS was to solve all of these issues by configuring a totally integrated combat system of systems, including the procurement of new sensors, weapons, C^3 hardware and software.

As is evident from the brief statement of the problem above and the A-109 required method of recommending a preferred solution, this program was a systems engineer's or analyst's dream. It involved virtually all of the disciplines and products of high-level systems analyses:

- The definition of alternatives at all levels—mission, system, subsystem, and technology
- The establishment of a high-level value structure that encompassed both warfighter and acquisition system requirements to guide the trade-offs and selection process
- The development of a formal requirements decomposition and tracking process
- The development of a formal system synthesis process that could service the definition of multiple systems and their associated derived requirements
- The development of high-level combat simulation and evaluation tools to define the performance of each alternative system construct across multiple missions in a variety of multimission combat scenarios

And beyond all that, the program required a way to take this multidimensional exploration of alternatives and present and explain, in understandable ways, what the important issues were with each alternative; why they were important; what was fundamental about these issues; and how, in the aggregate, across the domains of cost, risk, multimission effectiveness, operational ease, supportability, etc., they led to whatever preferred alternative our team was selecting. As I said above, a systems engineer's dream program.

Three industry teams won the concept definition contract, and I was the technical director for our team. Our program manager Howard Wing was a person who meshed well with my capabilities—while I was largely technical and generally impatient with management issues, he was very innovative in terms of managing a large complicated program such as SIRCS. He also had sufficient technical competence to understand the nuances in the ongoing technical work and how to make them best understandable to the customers in our interim program reviews. In thinking about how we could structure the system alternatives we were about to develop, we came up with the idea of "flavors." Rather than develop a number of system alternatives that might or might not result in significant differences or attributes, we decided to structure each alternative around a different flavor.

A flavor was to define, at the highest level, a unique—and, to the extent possible, orthogonal—system attribute. We defined flavors such as least complex operationally; highest integrated operational effectiveness across all three mission areas; lowest risk (which also turned out to be least development and quickest to initial operating capability); highest selectable performance in any individual mission area (i.e., if a mission area at a given time was the most stressing, its capabilities could be increased significantly at the expense of the others, while still maintaining a threshold performance level); and lowest life-cycle cost, including procurement, ship installation, manning, and logistics support. After one iteration of defining system constructs to best meet each of these flavors and going through a preliminary evaluation according to a value structure we had defined, we found that although each construct was indeed superior in the attribute for which it was designed to excel, there were sufficient similarities between the systems to allow a second iteration based on mixing, matching, and fine-tuning. This second iteration resulted in three distinct system constructs, maximizing the potential for either lowest life-cycle cost, highest performance, or least risk.

The use of initial flavors, and how it led to the learning, mixing, and matching to define the final flavors, turned out to be a big hit, with both internal personnel and the customers. For the customer community, the flavors formed a vehicle upon which to simplify the discussion of differences, attributes, and rationales within a very complex set of system-of-systems alternatives. Internally, the flavors allowed us to keep track at the highest level of what we were trying to accomplish within a given system, to provide an easily understandable framework upon which to conduct trade-offs and evaluate alternatives, and, most of all, to structure the work of myriad lower-level activities within a process that was at once reasonably unconstrained, open to as much innovation as any contributor could provide, and, most of all, convergent.

The three years that we were engaged in the SIRCS competition were among the most professionally pleasurable and rich in systems engineering and analysis learning of my entire career. Unfortunately, the program was cancelled at the conclusion of concept formulation. Circular A-109 had gone out of favor, the Navy's implementation of it in SIRCS had taken away some of the traditional roles of the Navy labs and constrained their influence on the program, and because of the nonstandard method of specifying capabilities across multiple, traditionally stovepiped warfare areas, SIRCS was not well supported in some segments of the Navy. Ultimately, Congressional politics and a seeming lack of Navy support killed the program.

However, using flavors stayed with me for the rest of my professional career, and I was able to use them successfully in a variety of future systems engineering activities. I have schooled young aspiring systems engineers in the use of flavors and have seen them used far more skillfully by these new generations than we had used them more than 30 years ago.

5.8 U.S. Army Air Defense and the Patriot System: The Evolution of Air Defense, Threats, Missions, and the International Security Environment

Throughout the 1970s, Raytheon had been involved in a series of system definition and development competitions for the U.S. Army, all focused on fielding a new generation of medium- and long-range ground-based air defense. The system was initially known as Field Army Ballistic Missile Defense System (FABMDS), then as Advanced Air Defense System for the 1970s (AADS-70),[2] then Surface-to-Air Missile–Development (SAM-D), and finally, after we won the advanced development contract, as Patriot. The system involved the first use of a phased array surveillance, tracking, and illumination multifunction radar; a very high-speed, high-acceleration, sampled-data homing guided missile; multirouted communications; and centralized, hierarchical software control of all system functions at both the battery and battalion levels.

Development of Patriot proceeded through the 1970s and early 1980s with ups and downs as in any major system development. After a vigorous test program, the first battalion of Patriots was fielded in West Germany in 1981. Just prior to that, in

2 FABMDS was initially conceived as having a tactical ballistic missile defense as well as an air defense capability. After the Anti-Ballistic Missile Treaty in 1972, the ballistic missile capability requirement in FABMDS was eliminated, and the program was renamed AADS-70. The tactical ballistic missile requirement was not reinserted into the system until the mid-1980s.

1979 as I recall, U.S. intelligence observed the Soviet Union testing improved versions of their Frog and Scud tactical ballistic missiles (TBM), the primary enhancement being greatly improved accuracy. These missiles became a major cause for concern to the United States,[3] and the Army Science Board (ASB) was asked to establish a task force to perform a quick study to examine near-term ways to counter this threat.

In conducting its system study, the ASB quickly determined that there were really only two potentially feasible ways to provide a near-term interim solution to the tactical ballistic missile problem:[4] either provide an adjunct to Patriot based on the subsystems contained in the Aegis system to provide the local advanced tactical ballistic missile defense, or modify Patriot itself to serve this function. To provide input on these options, the ASB invited both the Navy and the Army to come to the task force's next meeting to discuss these two options.

The Navy, in typical Navy style, appeared at the next meeting with a variety of flag officers, captains, and subject-matter experts from the Aegis project management office. The next three hours were spent essentially extolling the virtues of the Aegis system and the fact that it could be readily modified to do whatever was needed, and ending with the simple message, "Send money and we will do it." In the afternoon, an Army colonel and major appeared, and in a much shorter briefing, described Patriot and the recent fielding in Europe. The briefers ended with their own simple message, to the effect of, "We are up to our ears in getting Patriot fielded and producing missiles and fire units, so why don't you come back to us in a couple of years and we will see what we can do."

---

3 Patriot had been deployed in Europe as a part of NATO's flexible response doctrine. The essence of that doctrine was that NATO would employ its tactical nuclear retaliatory forces (dual-capable aircraft, cruise missiles, TBMs, and artillery) under a given set of conditions. The doctrine was purposely vague on what conditions might trigger the use of this option, but there was nothing vague in NATO's commitment to making its nuclear retaliatory forces survivable under any possible Warsaw Pact (WP) invasion scenario. The role of Patriot was to safeguard this option. Positioned both in the rear of then West Germany and on the political border between West and East Germany, WP bombers attempting to penetrate to the rear areas of NATO by overflying short- and medium-range air defenses would be severely attrited by the forward Patriots. This action would give NATO defensive counter air time to further engage WP aircraft before they got back to the rear area, in what was called the fighter engagement zone (FEZ). Whatever WP bombers managed to get through the FEZ would finally be met by the Patriots guarding the assets themselves. Thus, this three-layer zone of defense all but guaranteed the survival of the nuclear forces and the viability of NATO's local air superiority. A highly accurate TBM that had the potential to take out the Patriots threatened this entire strategy.

4 The ABM Treaty process had generated a cadre of arms-control scholars, both inside and outside the DoD, who were determined to root out any ballistic missile defense capability from U.S. air defenses. The earlier pre-Patriot FABMDS advanced tactical ballistic missile capability had fallen victim to this, at least in hindsight, shortsightedness. Even with the emergence of the former Soviet Union tactical ballistic missile threat to Europe, there was still significant debate about whether and to what degree to modify Patriot or the Navy's Aegis to establish a counter to this new threat.

That evening, Mike Fossier received a call from the ASB task force describing what had transpired and asking us and one or two other contractors to appear early the following week so that, in our case, they could get an answer as to modifying Patriot to have the ability to counter the emerging tactical ballistic missile threat in Europe. We were asked to stick to the technical possibilities—no marketing or programmatic material—so that the task force could make an informed assessment. Mike responded that we had never looked at this but that we would be happy to do what we could over the weekend and come down with at least the outline of a concept the following Tuesday.

Mike, another colleague, and I spent the next few days looking at possibilities. I raise this whole story in the context of systems engineering and analysis because it provides an excellent example of the role and value of back-of-the-envelope analysis against today's accepted paradigm in which answers can be provided only by employing very complicated, "validated" simulations. While these simulations may be valuable in providing detailed, high-precision answers to questions about very mature systems, they have virtually no role in providing quick-response answers to what-if feasibility questions. I personally feel that this ability to provide quick order-of-magnitude analyses has become more and more of a lost art in today's systems engineering circles. In his later years at the company, Mike also believed that the community as a whole was losing its ability to think through problems and do back-of-the-envelope systems analyses. He attributed this growing trend to the demise of the slide rule. Silly as this may sound at first, Mike had a credible rationale for his position. Since a slide rule provided only a two-digit (or if you had really good eyes, maybe three-digit) answer and no decimal point, you had to know without any aid except your own head where to place the decimal point. That restriction required you to understand the problem and the analysis sufficiently to know the order of magnitude of the answer. That understanding, in turn, required you to think through what you were doing. Unfortunately, the advent of the calculator, and later the computer, gave the impression of providing complete answers and mistakenly led the analyst away from having to think through the problem.

Over the weekend, the three of us worked the problem based on little more than distance equals velocity times time, the radar range equation, a half dozen basic Patriot radar and missile parameters, and a few threat parameters provided to us by the ASB task force, such as the radar cross section of the TBMs and their velocity profiles. We quickly determined where intercepts had to take place (a) if Patriot were just to protect itself and (b) if it were to protect both itself and the rear assets it was defending. We translated these two intercept regimes into the corresponding commit altitudes and radar detection ranges, and in turn to the required radar resources. The fact that the

Patriot radar was a phased array under software control and was powerful enough to support this mode of operation put the concept into the realm of feasibility. But as Mike often said, "We still need to prove that it can be 'feased.'"

Two issues needed to be addressed. Although the radar had the raw sensitivity to support the required detection and track ranges, it did not have sufficient resources to do this while at the same time performing every other function in its air defense mission. We found that a significant fraction of Patriot radar resources were going into providing long-range surveillance and that, because of the typical overlap in coverage at long range between the multiple radars in a Patriot battalion, we could have only two radars perform long-range search and the others perform TBM search to provide the required resources. All the units would do short- and medium-range air defense, and all would be capable of conducting TBM engagement. Since everything was under software control, the search allocations could be rotated by the commander to prevent preferential attacks.

The other issue that needed to be addressed was Patriot missile miss distance and lethality. Miss distance was not a major issue because both the Patriot missile guidance update rate and lateral acceleration were high, and the required changes in the guidance algorithms to accommodate the characteristics of a ballistic missile instead of a manned aircraft could be easily changed, since the algorithms resided in software on the ground. The appropriate aircraft or TBM guidance algorithms could be called up depending upon the type of target being engaged. More significant was the issue of lethality, given good miss distance. We came up with the following concept: make changes to the missile fuze to provide earlier detection for supporting the higher closing velocity, and to the blast-fragment warhead to make heavier, fewer fragments that would have the same lethality against aircraft as the current warhead but far superior capability against the more difficult ballistic missiles. Unfortunately, both of these concepts required implementing hardware changes to the missile, which would take significantly longer than implementing the software changes that all of the other modifications required.

We presented these findings to the ASB the following week, and when they asked what could be done quickly, we invented a two-phase program on the spot. The first phase would be to make only the software changes. The rationale was that software changes could be done quickly and would be sufficient for self-defense. Although the probability was not high that the Patriot missile armament would produce a hard kill against the TBMs, it would probably be sufficient, on an energy exchange basis, to deflect incoming missiles, spoil their accuracy, and significantly raise the probability of survival of the Patriot fire units themselves. The second phase would entail a few more software changes to enlarge the area of protection and, more fundamentally, hardware changes to the missile to provide the fuze and warhead modifications to increase lethality against TBMs. After an additional Defense Science Board study on the same issue, the concept was approved, and these two phases ultimately became Patriot Advanced TBM Capability 1 (PAC-1) and PAC-2. They were first fielded in 1988 and 1991 respectively, the latter during the Gulf War.

5.9 A New Kind of Analysis, and Learning How to Balance Breadth with Depth

In the years that followed, it sometimes became necessary to reinforce, both to the United States as well as our allies, the importance of Patriot to the security posture of the North Atlantic Treaty Organization (NATO) in Europe. It fell upon our group to provide the required analyses that would relate high- and medium-altitude air and missile defense to the higher-level Cold War security environment of the NATO alliance. This was a different kind of systems analysis from what we had typically done in the past, a broad operations analysis that was able to draw the connection between one aspect of warfighting capability and an entire defense posture. We had it within our own capabilities to measure the effectiveness of a given number of ground-based air defenses against a postulated Soviet air threat in terms of enemy sorties defeated, tons of enemy ordnance delivered, the probability of main air base closures, blue air sortie generation, etc.; however, we had no capability to go a step further and measure those results on a hypothetical ground war in western Europe. This dilemma and strategic deterrence dominated most of the security concerns of the 1980s prior to the fall of the Berlin Wall.

Rather than build the capability ourselves—which we understood would take a lot of time, people, and internal funding, and probably not be credible in the end—we contacted a couple of think tanks that did these kinds of analyses for a living and were the folks whom the Pentagon depended upon when they needed such work. After hiring a think tank to help us out, we found in our first attempts to work together that we spoke different languages. What a word meant to us was interpreted as something very different by them. A quick lesson learned was that words were important in these higher-level analyses, and it was imperative to take the time to make sure we were really communicating. This lesson was hardly new since we had learned it many times in earlier joint analyses—radar people talked a different language from missile people, software people from hardware people, command and control people from anybody else—but this lesson was easily forgotten and often had to be relearned.

We soon were producing charts of FLOT or FEBA[5] movement as a function of the level of NATO ground-based air defense in Europe. More importantly, we were starting to learn how to perform a newer, higher-level systems or operations analysis, that of campaign-level force-on-force war gaming. My own personal experience in systems analysis and engineering had broadened from the analysis of signal processing techniques, to radar system engineering, to other subsystem engineering, to weapon system engineering, to mission analyses, and now to analyses focused on the ability to carry out national security policy.

I also found it useful to think about the analyses serving these different levels as being characterized by a constant area, the dimensions of which were breadth and depth. If the breadth was small, then the depth (the degree of fidelity, the amount of detail, the inclusion of second- and third-order effects, etc.) could be very large. At the opposite end of the spectrum were the campaign analyses: very broad, but the characterization of the individual contributing elements done in relatively gross terms. Any ability to comprehend the results of these analyses, get a feel for whether they are right or wrong, perform back-of-the-envelope checks, do them in a reasonable amount of time, or explain the why's to someone else had better obey the constant area rule. As the availability of computer processing power has grown in recent years, I have seen a tendency to try to connect the lowest levels of detail to the highest levels of effects. On the basis of my experience over the years, I believe this tendency is a mistake. The issue is not available computing resources; the real issue is the analyst's ability to understand his or her computations and the results being produced. The problem must be addressed through a hierarchy of analyses in which the mapping of a lower-level capability onto a higher-level one doesn't exceed one or two levels, and the results at the higher-level are characterized as a few simplified parameters as an input to the next-level-up analyses. This hierarchy of analyses, of course, requires the systems engineers to really understand the problem they are trying to analyze or solve, and no computer in the world can substitute for this human understanding. Thinking that it can is a simple path to analytical disaster.

Mike Fossier retired in 1991, and I took over as the leader of our Advanced Systems Group. By this time, the group had expanded to about 20 engineers, small enough so that my nonexistent management skills did not present a problem, but large enough

5 FLOT (forward line of own troops) and FEBA (forward edge of battle area) were standard terms used during the Cold War to describe the battle line as a function of time between Warsaw Pact and NATO forces. The numerically superior WP forces would always drive NATO forces back to the Rhine River, but a measure of the quality of the NATO posture would be the time it took this to happen—long enough for massive reinforcements to arrive before the main operating bases were overrun.

so that we could perform fairly comprehensive, large-scale system studies internally or with only a relatively low level of support from the laboratories. Our group consisted entirely of systems engineers with a variety of backgrounds, including radar; aerodynamics; propulsion; guidance and control; missile seekers; command, control and communications; and operations analysis. All had an analytical bent, all had that instinctive sense of not getting hung up on squeezing the last tenth of a dB out of an analysis when it didn't matter, and nearly all had some hardware or software design experience from earlier in their careers. Both Mike and I had felt this design background was important because time spent actually designing and building things seemed to impart a lasting respect and understanding for reality regarding "wonderonium."

5.10 The End of the Cold War, New Analysis Challenges, and the Defense Science Board

Under Mike, we had become the go-to place for quick what-if analyses, feasibility assessments, internal red teaming, and fire fighting, particularly when issues of cost-effectiveness, system architecture, mission performance, or new system concepts were involved. This role continued and expanded somewhat during the 1990s, but three other increasingly significant changes put even greater emphasis on our systems engineering and analysis capabilities.

The first change was the growing impact of the end of the Cold War. The entire analysis community had become used to analyzing various aspects of a potential confrontation between the United States and the Soviet Union, and we all understood (or at least assumed we understood) Soviet capabilities, ways to model them against our strengths and weaknesses, the bipolar political environment, the two sides' goals and objectives, and all of the other aspects of the Cold War that we had dealt with for 40 years. Suddenly, all of this disappeared. We were confronted with a totally new international security environment that contained a variety of bad guys we didn't understand, capabilities we didn't understand, and, most of all, objectives that had little to do with defeating us militarily but a lot to do with damaging us politically, creating chaos, attracting others to their cause, and all kinds of other things with which we had little familiarity. New rules, new objectives, new measures, and new analysis techniques had to be found (and still are being found), and all of these greatly impacted systems thinkers and analysts.

The second significant change was the proliferation of tactical ballistic missiles into Third World countries and the use of these missiles in the Gulf War. These circumstances significantly increased the international market for air and missile defense and early warning. It fell upon our group, as the company attempted to sell Patriot and large early-warning radars into this emerging market, to make the technical cost-effectiveness case for these products. Here, in the very early stages of these campaigns, there was no existing requirement; in order to establish one, a case had to be made on both the political and the military level as to why the cash outflow to purchase such expensive systems was worthwhile in terms of the value received. Trying to express this case in traditional weapons system attributes—such as engagement range, target handling capability, probability of kill, radar surveillance volume and range—meant little. Although these attributes were interesting to some of the junior, more technical military personnel, to senior military personnel or to the political leadership the rationale had to be made at a much higher level. That higher level required understanding the national security concerns of the particular country involved and any differences (there always were a few) between what was felt to be important by the military and what was important to the political leadership. This situation was often complicated by party differences among various factions within the political structure of the country.

We would read as much as we could find on the issues within each country, talk to people who had spent time there in various capacities, and consult political and military analysts. On the basis of what we learned, we put together briefings that attempted to translate our systems' military capabilities into terms that were of value to those to whom we were presenting. Often our initial understanding was wrong, and what we thought a particular country (or element of that country) valued turned out to be something of little concern to them. However logical it seemed to us to be worried about something, it was not what they really cared about, but something else was. Depending upon the country involved, some of the issues that we learned to address were the preservation of economic value; the weighted percentage (the weighting factor was population density) of civilian population that fell under an umbrella of protection; the defensive and offensive counter air sortie generation as a function of enemy attack time; the strategic deterrence of air attack by a totally different adversary from what we had assumed; and the time available for political decision making and de-escalation before a massive response became politically mandated. Many of these concerns were new to us and required significant learning and analysis capabilities to address. Once again, we found analytical ways to determine general trends, to iterate answers and update our simple models as we learned more from our potential customers, and to convey our results in easily understandable but credibly supportable ways.

In most cases, given that we had done a reasonable job of understanding a particular country's interests, expressing things in those terms, and supporting our arguments with technical facts, we found people were willing to explain where our assumptions were wrong and what the right issues were. This conversation would lead to another iteration of the briefings and arguments, and ultimately we were able to put together a convincing case. Thus, although collectively we had the fundamental systems engineering skills to span the spectrum ranging from system-level technical details to warfighting capabilities to national security significance, the idea that one size doesn't fit all, that each campaign had to start with a reasonably good understanding of what made each country tick, and that listening was important was reinforced each time we ventured into a new endeavor.

The third activity that broadened our learning and deepened our systems engineering capabilities was my own growing personal involvement with the Defense Science Board (DSB).[6] In 1990, I was invited to participate as a consultant on a broad DSB summer study involving the fundamental changes in the U.S. national security environment brought on by the fall of the Soviet Union, and what those changes might mean to the Defense Department. The DSB participation was a whole new experience in many ways: the intelligence of the other participants, the breadth and depth of their collective as well as individual knowledge, the level of the issues the summer study was dealing with, and the interesting perspectives that were provided by a forum in which senior military, senior industrial leaders, and well-seasoned academics worked together to produce a product. The experience was also a great lesson in how to run a major study, particularly one in which a hundred Type-A personalities, all used to getting their own way, were herded together in a combined effort to produce, if not a consensus view, one that at least everyone could more or less agree to without too many conflicting opinions.

As I became increasingly involved in more DSB task forces and summer studies, I was able to bring some of the analysis skills, perspectives, study management, and a host of other lessons learned back to our Advanced Systems Group and inculcate the more senior members with what I had been exposed to. I also began to notice, however, the impact that this exposure was having on the more junior members of our organization. Their perspectives broadened; they learned how to assess the level

6 The DSB is a high-level advisory board to the DoD that examines broad technical issues of potentially high impact that cut across traditional DoD organizational lines. Its studies are requested by senior levels within the Office of the Secretary of Defense (OSD) and fall into two categories: task forces and summer studies. Task forces are typically small groups of about 10 participants focused on a specific issue and involve monthly two- or three-day meetings over 6 to 12 months. Summer studies involve a much larger group of participants, deal with much broader issues, generally last about 9 months, and spend the one or two weeks collocated at an offsite, the conclusion of which is an outbrief of the results to senior DoD and other agency personnel. Membership on these studies is made up of DSB members and non-members, the latter appointed as consultants for the duration of the study. Details can be found on the DSB website at http://www.acq.osd.mil/dsb/.

of information they needed to arrive at a credible, useful assessment; they began to ask the right questions, both of themselves and their colleagues; they learned how to create briefings focused at the right level for what they were trying to accomplish; and in general, they exhibited very rapid professional growth. The key to this rapid learning was the junior members' immersion within a group of more senior systems engineers who were operating at a level above that generally found in industry, where systems engineering activities tend to focus on a particular product.

The rapid growth of these young systems engineers was noticed by senior management as well. We decided to conduct an experiment. Each year, we selected two promising young systems engineers, removed them from the System Engineering Lab, and put them into my group for a period of two years as a sort of apprenticeship. In the end, everyone benefited: the company because a new generation of very capable systems engineers was developed who later took over senior technical leadership positions in the company; the DoD because more useful system-level expertise was brought to bear on the department's more pressing problems through various program offices; the DSB and other Service science boards; and, of course, the individuals themselves as their maturity, systems engineering capabilities, and overall competency increased, and they assumed increasingly important levels of responsibility.

5.11 A Few Concluding Tips for the New Generation of Systems Engineers and Analysts

I retired in the summer of 2000, although I continue my association with the DSB, now serving as a Senior Fellow and participating in many ongoing studies. I also continue my association with industry as a consultant, helping to sort out some of the more complex systems engineering issues of the day. Looking back on 50-plus years of systems engineering and systems and operations analysis, I cannot resist this opportunity to share a handful of lessons learned with any aspiring young system engineers who have the inclination and patience to keep reading:

- Understand and think through whatever problem you are working on sufficiently for you to come up with a ballpark answer without resorting to lots of computational resources. In order of priority, the tools you should employ are your fingers, then an abacus, then a hand calculator, then a spreadsheet, and as a last resort, MathWorks. Stay away from large-scale simulations, at least initially.
- Avoid the temptation to provide six-digit answers to problems with one-digit (at best) inputs and assumptions. At a minimum, those answers are unnecessary

and of absolutely no value, and worse, among seasoned analysts and engineers, such overkill will be a red flag on your credibility and will highlight that "you really don't understand."

- Don't be satisfied with any answer you get to a problem without first asking yourself a few questions:
  - Does it make sense on first principles? Can I explain the *why* behind this?
  - If I alter the inputs to a trivial or extreme case, do the answers I get seem logical (passing the *reductio ad absurdum* inspection)? Do the trend lines make sense?
  - What could screw up this answer? Which of the inputs, if only slightly different, could lead to a very different answer? Does my analysis sit on the edge of a cliff? Could a different way of employing a threat drive to a very different conclusion? In effect, be your own red team. And after you are done, ask someone else whose intellect and understanding you respect to sit down and destroy your approach or conclusions (it only hurts for a little while).
- Make sure you have treated all of the important, driving parameters, but have not wasted time, attention, and effort on trying to accommodate all of the other less important parameters in the interest of completeness. This, of course, is simply another dimension of understanding the problem, both to enable you to separate the wheat from the chaff as well as to explain to others why you have safely ignored certain second-order issues.
- Keep your briefings, explanations, and reports simple and focused on the important issues. You need to have established the technical support for your findings, but they needn't be so intertwined with the important issues that the major thrusts of your arguments get lost in the noise. Your ability to verbalize that supporting data when challenged (and to pull it out if absolutely necessary) will more than establish the credibility of what you have done. Ditto the use of appendices in reports—most people won't read them, but for those who do, you will have established that you have done your homework. Good up-front executive summaries are great.
- Be patient with those who may not have your degree of knowledge, analytical skill, or ability to see the whole through the parts. Learn to make common-sense explanations analogous to things most folks understand: sports, driving a car, painting a billboard, whatever. If you have thought your analysis through, there is usually some analogy that will help in getting across what you

are trying to say. But learn to do this without being patronizing—remember that whoever it is that you are briefing probably knows something that you can't easily comprehend, at least the first one or two times that you hear it.

- Most of all, as your expertise as a systems engineer/analyst increases and you get assigned to more and more challenging problems, bring along one or two promising young engineers, just as you were not very many years ago. Let them learn from what you are doing, let them participate in formulating answers, wean them away from their electronic aids, and teach them to think. Soon you will have created another you. There is no greater reward.

About the Author

Robert Stein is a private consultant to the defense industry and the U.S. government. Prior to his retirement in June 2000, he was a board-elected Vice President and Officer of Raytheon Company and managed Raytheon's Advanced Systems Office. He was responsible for the formulation and implementation of advanced systems and concepts for current and future Raytheon government product lines, covering the full spectrum of modern missiles, radar, electro-optical sensors, C^3I, integrated systems, and associated technologies. He also oversaw the allocation and expenditure of independent R&D funding and activities for the company.

Mr. Stein joined Raytheon in 1958 and performed a variety of engineering functions over the following 42 years, developing a wide range of systems in radar, air defense, missile defense, and command and control. He led concept formulation studies in strategic and tactical defense systems.

Mr. Stein is a Senior Fellow of the Defense Science Board and has served on more than 20 DSB summer studies and task forces dating back to 1990. He has performed special consulting assignments for the MITRE Corporation, DARPA, the Missile Defense Agency, and Lincoln Laboratory.

Mr. Stein performed undergraduate studies in electrical engineering at MIT and graduate studies at MIT and Boston University in mathematical physics. He holds a patent in multi-beam radar antenna techniques, has published numerous articles on defense technology and related policy issues, and has taught a variety of courses on radar and information theory.

In 1992, Raytheon awarded Mr. Stein the Thomas L. Phillips Award for Excellence in Technology, the company's highest recognition for technical achievement.

6

Systems Analysis and Red Teaming

Aryeh Feder

> As a scientist or engineer, you were trained in the details;
> as a systems analyst, you avoid the details as much as you can.

6.1 Introduction

I joined Lincoln Laboratory in 1997 after completing my PhD in physics at Harvard University. My thesis had been in experimental physics, specifically optical studies of liquid surfaces. While I had enjoyed my time in grad school, by the time I had finished, I had come to the conclusion that experimentation was not for me. When I interviewed for a job at Lincoln Laboratory, I discovered a career that I never knew existed: systems analysis. This job combined my favorite aspects of physics—analyzing and solving problems—with an opportunity to make important contributions to important problems.

My initial work in air vehicle survivability offered the opportunity to learn the ropes from some of the best systems analysts in the country. I wound up leading an analysis group for some seven years, and then did a two-year tour in Washington, D.C., as the chief scientist for the Special Programs Office, Undersecretary of the Air Force for Acquisitions. I returned to Lincoln Laboratory in 2011 to assume leadership of a group that was analyzing system architectures for missile and air defense systems. My hope is that I may have some useful advice for those starting out in systems analysis.

6.2 What Is Systems Analysis?

As is evident from the variety of perspectives discussed in this book, there are perhaps as many definitions of systems analysis as there are practitioners who consider themselves to be systems analysts. At its core, however, systems analysis is a methodology for understanding the basic trade-offs that drive a system's performance. Modern defense systems are highly complex integrated systems that interact in complex ways with each other and with the environment. However, for many important analyses relating to these systems, creating a high-fidelity, detailed model of these systems and all of their interactions is a poor approach, for reasons discussed later in this chapter. Luckily, despite the complexity of the systems being analyzed, many important questions can be addressed with much simpler models that focus on the physics relevant for a particular question.

The goal of this approach to systems analysis is not to answer specific questions about what the exact system response will be in a particular situation. Rather, the goal is to understand what the important drivers of system performance are in order to enable decision makers to make informed decisions. Through my experience, I have developed the following basic rules for systems analysis:

1. Start with the big picture view. It is important to understand the context in which a particular question is being asked.

2. Focus on the key physics that will drive the answer to the question. Many other effects may be critical to the function of the system you are assessing but not to the particular question being addressed; your model doesn't need to encompass all the effects.

3. Model only to the degree of fidelity necessary. The more you know about a system, the more tempting it is to include those details in your modeling. Try to ignore the temptation to be specific when a generic model will capture the important effects.

4. Keep the big picture view. Ultimately, what is important is not the model of the particular physics that you are considering, but how that physics drives the final system performance. Make sure to put the results back in that big picture context.

This systems analysis methodology can be applied to a wide range of problems. One important class of problems is system architecture trades. These system trades range from the very large scale (e.g., what components do I need to create an effective ballistic missile defense system?) to the smaller scale (e.g., what are my options for detecting radio frequency (RF)–triggered improvised explosive devices?). Systems analysis enables researchers to perform the key trades in identifying promising architectures that can then be assessed in greater detail. Another important class of systems analysis problems is requirements analysis (e.g., what power should I require for my new jammer?).

However, the focus of this chapter is on the application of systems analysis methodology to the task of *red teaming*. Generally speaking, the name red team can be applied to any group given the task of looking at a system from the perspective of an adversary, specifically attempting to find weaknesses that an adversary may exploit. Red teaming can be used retrospectively to find and fix problems in systems that have already been developed. For example, government or corporate entities concerned about the security of their computer networks will often engage a red team that attempts to hack into that system as a way of uncovering vulnerabilities that the system designers overlooked. Red teaming can also be used prospectively to judge the advisability of investing large sums in a particular advanced technology or system. In particular, red teaming can offer a hopefully unbiased assessment of the potential responses of our adversaries and the impact that those responses may have on the technology the United States is considering. Clearly, the goal is to avoid investing in technologies that have simple and inexpensive counters, and to focus on technologies that appear robust to likely threat responses. A prominent example of this concept is the Air Force Red Team, which has been advising senior Air Force and Department of Defense leadership on air vehicle survivability and other critical issues for more than 30 years.

6.3 A Little Bit of Red Team History

The Air Force Red Team was formed in the early 1980s to support U.S. development of low-observable air vehicles commonly referred to as stealth aircraft. Survivability of aircraft has always been a concern of the military, and this concern grew as air defenses became increasingly capable. During the Cold War, the Soviet Union had assembled a formidable complex of air defenses with surface-to-air missile systems and advanced fighters, both of which were largely dependent on the detection of aircraft by long-range radars. The traditional response to these threat radars was the development and fielding of ever more sophisticated electronic countermeasures. This response would then motivate adversarial radar designers to develop counter-countermeasures, and the countermeasure/counter-countermeasure cycle continued to repeat.

While the development of electronic countermeasures remains important, in the late 1970s the United States realized that substantially reducing the radar observables of an aircraft was technologically feasible and potentially more robust. Thus began a large, extensive, and costly effort to develop stealth aircraft in a secret fashion. The immediate concerns were, could it be made to work, just how well could it work, how well did it have to work, and how would a competent adversary react? These last two questions provided fertile ground for the need for a very capable Air Force Red Team.

Note that stealth aircraft are called low observable, not unobservable. There is an inclination for the uninformed to equate stealth with invisibility. This inclination stirred the creative minds of scientists and technologists in the United States and elsewhere to propose a wide variety of unconventional techniques to detect and track these supposedly invisible vehicles. So the early Red Team had a jump start in dreaming up possible adversary reactions to our stealthy air vehicles. The team's main effort would be to run down a wide variety of evolutions or improvements in conventional-style air defenses. The team also had to cast a broad net to cover some decidedly "kooky" schemes. Ultimately, the Red Team's job was to assess approaches that might counter our significant investment in stealth aircraft, and to determine the likely survivability of our aircraft against those defenses.

The Red Team that was formed at Lincoln Laboratory grew to more than 200 staff members who enlisted substantial consultation from other entities in the air vehicle community. It also gave birth to a number of similar efforts at the Laboratory that feature the analysis style discussed here.

6.3.1 Typical Red Team Questions

Generally speaking, a red team might ask a series of questions such as the following:

- What are the key gaps in my system or capability that an adversary might exploit?
- What are the countermeasures that an adversary could develop to exploit those gaps, and how effective would those countermeasures be?
- How difficult would it be for an adversary to implement the countermeasure? Would the countermeasure be a 10-line software modification, or would it require building a new system from the ground up?
- How much detailed information about the U.S. system does the adversary require to make the countermeasure effective? Does the adversary just need to know our system's general concept, or would the countermeasure depend on specifics that are classified or otherwise difficult to acquire?
- How difficult would it be for the United States to counter those countermeasures that might be effective? How much information would be required about what the adversary is doing to effectively counter the countermeasure?

- What are the costs to the adversary, either monetary or in reduced system performance, to employ the countermeasure?

Given that our adversaries have finite resources (as do we), one would expect that they would invest in the technologies that are likely to give them the best bang for their buck. That is, if there are countermeasures that analysis indicates are likely to be effective and easy to implement, that do not require detailed classified knowledge of how our systems work, that would be difficult for the United States to counter, and that do not have significant cost in money or performance for the adversary, we can be fairly certain that an intelligent adversary will prioritize the development of those technologies over other less attractive options.

The Air Force Red Team is much more than a systems analysis group to explore these questions. Much of the Red Team's work involves the development and testing of prototypes of potential future threat systems. Importantly, these prototypes are not based solely on intelligence estimates of what technologies our adversaries are investing in. Instead, the red team process uses systems analysis to determine the technologies that would result in the most pertinent threat capabilities, and then develops prototypes that could be used to test the real-world impact of those technologies in tactically relevant scenarios. Prototyping and testing alone, however, can rarely be used to explore the full impact of a potential threat in a broad range of tactical situations. Therefore, the results of the testing are used to validate the systems analysis models, which can then be extended to assess a range of relevant scenarios. This integrated program of systems analysis, prototyping, and instrumented testing is the core of the highly successful Air Force Red Team. The real benefit of the analysis in this context is to sort through the large number of potential threat responses and narrow down to the few that are the most important. Those few can then be the subjects of more intensive prototyping and testing campaigns to understand the full impact that they might have on our systems.

6.3.2 An Illustrative Red Team Problem

An example of a red team analysis is useful here. The Air Force Red Team advises Air Force senior leadership on which technologies to invest in to make it most difficult for an adversary's air defenses to engage and shoot down U.S. aircraft. When we consider aircraft survivability, our favorite chart is the "kill chain" illustrated in Figure 6.1. Engagement of an aircraft starts with some sort of surveillance to provide as complete a picture as possible of the activity within an airspace; therefore, surveillance sensors provide long-range detection of large numbers of targets, but with relatively low-quality track of each target. The air picture from multiple surveillance sensors

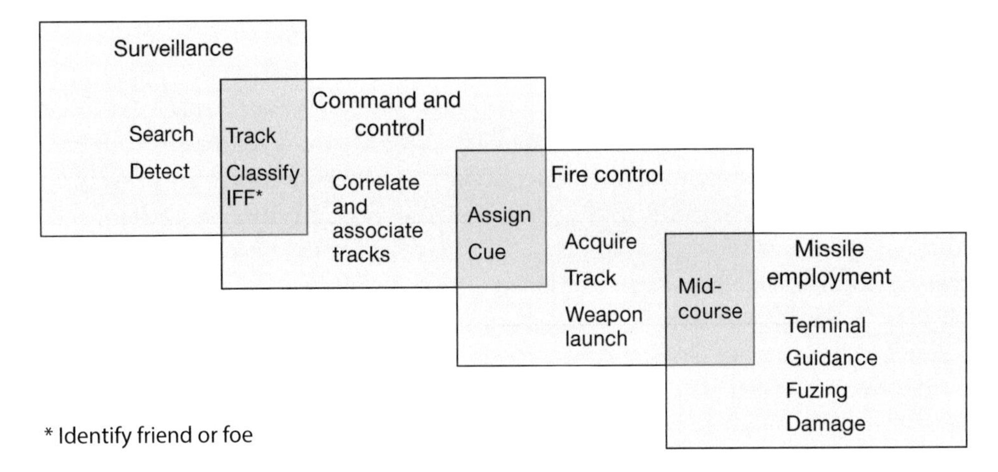

Figure 6.1 Air defense kill chain.

can be fused to allow command-and-control decisions, such as the identification of friendly and unfriendly aircraft and the assignment of a target to a particular engagement asset. This assignment consists of a cue to the so-called fire-control sensor whose role is to track a small number of targets (perhaps only one), but to track them with sufficient quality to guide a missile to intercept. The missile will often have a seeker that homes in on the target to bring about the end of the engagement, which is the fuzing and detonation of the missile warhead.

For each air defense element (surveillance, fire control, missile seeker, and fuze), the air defense can use a variety of sensor technologies, including active or passive radar (radio frequency), infrared, or a range of other less conventional sensors. The unconventional approaches proposed as counter-stealth technologies may run the gamut from acoustics to cosmic rays to gravimeters; however, in each case, it is critical to consider not just whether or not a technology can theoretically detect an aircraft, but whether it meaningfully contributes to the adversary's kill chain. For a number of good physics reasons, air defenses rely primarily on radar and infrared sensors. Compared to infrared, radar has the advantage of offering long-range detection with minimal interference from weather or other environmental factors. It is no accident that the most important long-range surveillance and fire-control systems continue to be active radars. One disadvantage of radar systems is that they are potentially susceptible to jamming, also called electronic attack. Therefore, for as long as air defenses have invested in radars, aircraft designers have invested in electronic-attack countermeasures to those radars, and, in turn, radar designers have invested in electronic-protection counters to those countermeasures in an endless cycle. The key on each side of the cycle is to invest in technologies that are as robust as possible to the other side's likely response.

One class of electronic attack that the United States has invested in is called a towed decoy, illustrated in Figure 6.2. Towed decoys are countermeasures to radar-guided missiles, whether those missiles are guided by a fire-control radar or by a missile seeker. The basic concept of the towed decoy is to detect the radar energy, and to amplify and rebroadcast that energy to the radar, thereby providing a more attractive target to the radar than the aircraft itself. Because our adversaries are well aware that the United States has invested in and deployed towed decoys, it is safe to assume that they are investing in electronic-protection countermeasures, and that the United States should be investing in counters to those countermeasures. However, there are a wide range of countermeasures that our adversaries may pursue, and it is the role of the Air Force Red Team to assess those options from the adversary's perspective to prioritize the ones the United States should be investing in countering.

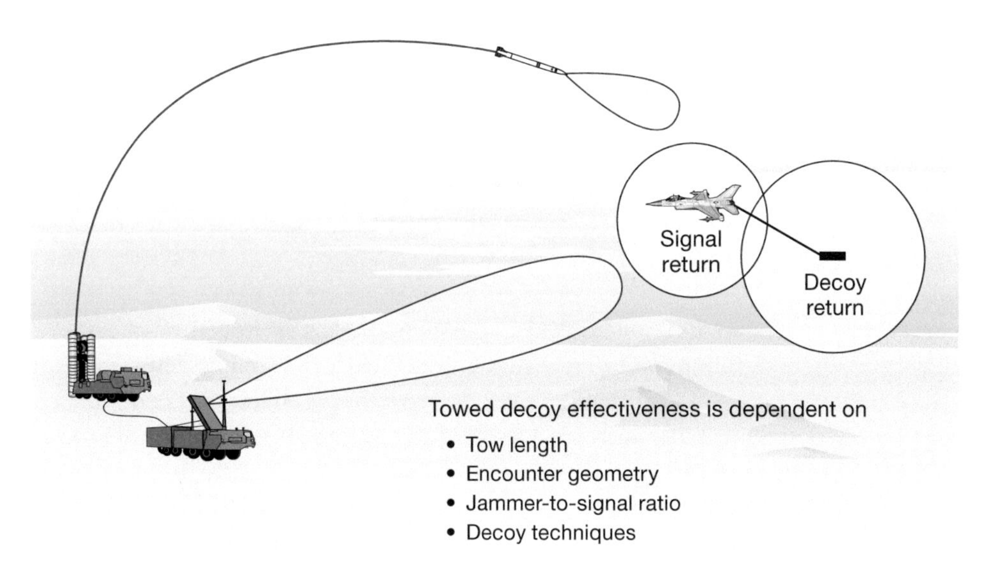

Figure 6.2 Towed decoy concept.

6.4 How Do You Do Systems Analysis?

The first and most important step in performing systems analysis is to understand the problem you are working on. If you are a new analyst, fresh out of school, chances are you are working on a problem that is largely or completely unrelated to anything you've done in the past. Take the time to learn about whatever it is that you don't already know. If you are an employee of a large high-technology organization, you're in luck; a national expert on the topic is probably working down the hall or in the next building. However, as you learn about the problem, remember that you are a systems analyst not a system designer. It is important to understand the details, but don't lose track of the big picture. For example, if you are assessing how well an advanced signal processing technique might allow a radar to distinguish between an aircraft and its towed decoy, it is important to understand the signal processing concept and the properties of the aircraft and decoy signal that determine how effective the technique may be. It may be necessary to code up a version of the signal processing technique and to feed it simulated data to understand its performance. Your job is not to squeeze every last bit of performance from that algorithm but to determine whether or not an adversary would likely use that technique, and if so, what the potential U.S. responses are. For this reason, the best systems analysts are people who enjoy constantly broadening their view and learning new things, but not getting lost in the details. Once in a while, some details really count. The hard part is to identify those cases without constantly getting lost in the details.

After you understand the problem that you're working on, figure out what physics, engineering, signal processing, or phenomenology is going to drive the answer to the question. For example, anti-aircraft missiles are complex systems that interact with many other complex systems. A high-fidelity model of an intercept would take into account numerous details, from the target trajectory and radar cross section, to the missile aerodynamics, seeker detection, countermeasure detection logic, signal processing, and so on. However, most aspects of the problem are of relatively small consequence for a particular question. For example, consider an analysis of whether a particular advanced signal processing technique will allow a missile seeker to distinguish between the radar return from the aircraft and the radar return from the decoy. Answering that question requires understanding radar and signal processing, as well as key aspects of the decoy, e.g., what power it transmits and what modulation, if any, it adds to the received radar signal. On the other hand, it is relatively unimportant to understand the details of the missile-guidance algorithm or detailed aerodynamics; one can assume that the missile will go where the seeker wants it to go. The key issue is whether or not the decoy can fool the seeker. Conversely, assessing the ability of a missile to intercept an aircraft without a towed decoy that is performing a high G last-second maneuver is almost the exact opposite; modeling the detailed seeker performance is unnecessary (with a few exceptions, you can assume that the seeker will do its job of tracking the target), but the target maneuver and detailed missile aerodynamics become much more important.

Once you've identified the key physics of the problem, build a model that captures the physics. If you are part of a group that has been working on similar problems for a long time, modify an existing model to capture the relevant physics. To the extent possible, the model should be a generic instead of a specific representation of a particular system. It is much more valuable to understand the performance over the relevant parameter space and to highlight where, approximately, a particular system lies in that parameter space than it is to create a detailed model of a specific system. Performing the modeling in this way allows you to understand the sensitivities to the key parameters and to the assumptions that you've made. It will probably even offer insight into why a system was designed in a particular way. Conversely, if your model indicates that a simple change in a system would dramatically improve its performance, this indication may forewarn that your model is not capturing some important effects (the system designer probably knows more about the details than you do).

Ultimately, however, your sponsors are not looking for the detailed results of your seeker or aerodynamics model; they want to know what the survivability of their aircraft is and what they can do to improve it. Therefore, after the detailed modeling of the key part of the problem, it is necessary to get back to the big picture by modeling the noncritical aspects of the problem with a relatively lower degree of fidelity. In theory, of course, it's possible to build a more complicated model that captures more than just the most important physics in the problem. In the missile example given above, one could create a model that captures both the detailed seeker signal processing and the full missile aerodynamics. However, the cost of adding extraneous details to the model is the gradual loss of transparency into the modeling, which is one of the key benefits of systems analysis. There are plenty of high-fidelity models out there, and while they may serve an important function, they are not systems analysis tools. The crucial part of systems analysis is maintaining your ability to apply your physics intuition to the problem you are addressing, ideally building test cases that are amenable to back-of-the-envelope calculations. That way, when the model gives results that are not what you expect, you can figure out if the modeling is flawed or if the physics is more subtle than you gave it credit for (this second possibility is the more interesting challenge!). The danger of high-fidelity models is that you lose the ability to effectively apply your intuition to the problem; you never want to defend your results by uttering the systems analyst's death words: "I'm not sure why it came out that way, but that's what the model said."

The final part of the process is one emphasized by all the authors of this book: figuring out what the story is that you are trying to tell and building the briefing to tell that story in a concise, compelling way. It is impossible to overestimate the importance of the story. Even after you get the technical analysis correct, your analysis won't have an impact unless you can convey the analysis in a clear story line that your customer can understand. Unfortunately, there is no simple recipe for building and presenting a quality briefing, but any organization performing high-quality systems analysis will contain role models who illustrate this talent in their briefings and who can help you determine the story to tell and the most compelling way to tell it.

6.5 The Important Role of Testing

Systems analysis is a very powerful and valuable tool. However, a systems analysis is rarely an end in and of itself. In particular, effective systems analysis requires a close connection to real-world testing. A systems analyst should always recognize that while modeling is important to quickly perform high-level system trades, modeling tools should always be viewed with a healthy dose of skepticism until (and even after) they have been validated with test data. This piece of advice does not mean that a lot of valuable work cannot be performed with unvalidated models; however, the real world often imposes constraints that may not be incorporated into a first-order model. In other words, you might be fairly certain that you have captured the important physics of a problem, but until you have tested and validated your findings, you should be open to the possibility that the real world is more complicated, and interesting, than you gave it credit for. In addition, testing is sometimes required as a prerequisite to effective modeling. In particular, many problems are not easily reduced to a first-order physics problem; it may well be necessary to create and execute a testing program to understand the phenomenology behind a problem in order to determine the best way to model the relevant system.

It is for this reason that the Air Force Red Team relies on a well-integrated process of systems analysis, capabilities-based threat prototyping, and instrumented testing. Each of the elements of the program contributes critically to the overall Red Team mission, and each is reliant on the other elements of the program. Systems analysis can serve to identify which are the critical technologies that should be included in a prototype of a future threat, and it can be used to help determine the critical parts of the parameter space on which to devote limited test resources. Conversely, prototyping and testing provide the real-world reality check to validate the systems analysis predictions, as well as provide the phenomenological data necessary to develop some systems analysis models. (Not unimportantly, demonstrations with actual threat prototypes are far more convincing to senior leadership than curves on a PowerPoint slide.) Highly instrumented testing (rather than stunt-like demonstrations) is critical because it provides the data to determine not only what happened in a test, but why it happened. Ultimately, the combination of all three of these elements has made the Air Force Red Team a model whose value is recognized across the Department of Defense.

About the Author

Aryeh Feder joined MIT Lincoln Laboratory after receiving his PhD degree in physics from Harvard University in 1997. As a staff member in the Systems and Analysis Group of the Tactical Systems Division, he performed analysis on a wide range of topics relating to air vehicle survivability, including assessments of a variety of radar and infrared sensor systems' performance versus U.S. Air Force aircraft, and the impact of current and potential future electronic-attack and electronic-protection techniques on these aircraft. As a group leader in the Systems and Analysis Group from 2002 until 2009, he was responsible for leading the analysis portion of the Laboratory's Air Force Red Team Air Vehicle Survivability Evaluation program, which provides system-level assessments to senior Air Force and Department of Defense leadership on a wide range of topics, including aircraft survivability, electronic attack and electronic protection, and a variety of Air Force special programs.

In 2009, Dr. Feder accepted a two-year Intergovernmental Personnel Act assignment as the chief scientist for the Special Programs Office, Undersecretary of the Air Force for Acquisitions. In that role, he assessed the technical feasibility of current and proposed Air Force special programs, identified critical capability gaps, and defined roadmaps for technology development to close those gaps. He also served on several high-level Office of the Secretary of Defense panels addressing issues related to advanced aircraft survivability and electronic warfare. In September 2011, he returned to Lincoln Laboratory and now leads the Systems and Architectures Group in the Air and Missile Defense Technology Division. He is responsible for systems analysis of a broad range of problems, from ballistic missile defense architectures to assessments of advanced technologies for U.S. air defense systems.

7

Blue Teaming and Telling a Good Story

Robert G. Atkins

7.1 Introduction

Systems analysis is probably one of those things best learned via apprenticeship—years working side by side with a master of the craft. Early in my career at Lincoln Laboratory, I was very fortunate to "study with" some truly great masters of this craft. I owe them most of what I know about it today. I didn't set out to become a systems analyst—I'm not sure anyone does, or that very many college undergrads even could tell you what a systems analyst does. Academic training, particularly the process of earning an advanced degree, tends to drive students deeper and narrower. I was no exception, and despite many years as a cooperative education student and summer employee of the Laboratory, when I started as a staff member, I can recall the challenge of balancing that pursuit of the details with a new system-level view. Eventually, the systems approach to problems became quite contagious; the opportunity to have grander impact won me over.

I had the opportunity in the first decade of my career to work a great range of systems analysis problems spanning air defense, ballistic missile defense, surface surveillance, and homeland protection. Later, I became involved with rapidly developing solutions to critical problems. The goals of these rapid efforts were to identify workable solutions to vexing military problems and to deploy these solutions in months versus the classical DoD acquisition cycle often requiring many years. It was in these efforts that we developed a set of approaches that collectively became known as "blue teaming." This blue-team process is complementary to that of red teaming, described in the previous chapter. While red-team analysis seeks to uncover vulnerabilities and shortfalls in defense-related systems, blue teaming seeks to develop engineering solutions to existing defense challenges and problems. It accomplishes this goal through an open process that emphasizes technology-agnostic innovation, modeling, and critical measurements to initiate and validate solution concepts. At the center of this process, of course, is a healthy dose of systems analysis. That analysis is needed not only to reach the solution but to present that solution in a clear, compelling manner—to tell a great story.

7.2 The Good Story

Everyone loves a good story. It's hard not to recall a few good stories that stay with us, perhaps even dating back as far as early childhood. For some of us, this memory might be a particularly gripping ghost story, punctuated by a spitting campfire, and ending with a clever twist worthy of the suspense the teller has spun and held us with from the opening words. For some of us, there was a particular book, the one we couldn't put down and with which we stayed curled on the couch late into the night to read just one more chapter and then another. For others of us, it might be a story told at a family gathering by that certain relative—we all know the one—who seems to have an endless supply of entertaining stories from an overly adventurous life. Like everyone else, I love a good story, but for me some of the most memorable stories are of a different kind. I love a good briefing.

Hopefully, everyone has had opportunities to experience this kind of good story as well. Not just your everyday well-constructed and well-delivered briefing—what one might call the acceptably good briefing—but the truly exceptional one. It is the one that captivates us from the first slide and keeps us asking the next question, only to advance to the next slide and answer it. It is the one that takes a problem like a hopelessly tangled coil of rope, and in front of our eyes magically untwists it to reveal a linear answer in a way that makes it look obvious. It is the one that invents a new option that none of us had thought of, and rescues us from compromise or from the dark void of no solution.

It is shocking how infrequently we find such a story. It is particularly amazing when one realizes how significant a role these stories can play in shaping the course of the businesses in which we all work. Briefings are about selling an idea or a product, about shaping the direction a venture will take. For most of us, gone are the days of high-school debate team with little more than a trophy and a bit of team honor on the line. Decisions move large sums of money, to be made or lost. Decisions win and lose wars. And stories can drive those decisions.

In remembering good stories experienced in the past, it is easy to confuse a good storyteller with a good story, or to give the storyteller most of the credit for making the experience so memorable. Clearly the storyteller matters. Charisma and connection to the audience—what some would call presence—is important. But although a poor storyteller can ruin a good story, even the best storyteller can do little more than a mediocre performance with a poor story. Briefings are no different. A poor delivery can slaughter a good story, but even the best briefers, performing on their best day, can only sell a bad story so far.

So why do so many briefings fail? Why are so many unable to rise to the level of a good story? The most serious failure occurs when the answer the briefing delivers is simply not right. The error is frequently either obvious to the audience, or at least apparent enough that something about the story doesn't seem quite right. Although we might all hope that such a gross failure would be caught in advance and corrected, there are several reasons why many times it is not. First, of course, it is possible that the briefer set out to deliver the right story, but the process of developing it was flawed by incorrect analysis or faulty data. This failure usually has less to do with the method of analysis and more to do with the inevitable assumptions going into that analysis. Particularly when the briefer doesn't understand the problem well (or the audience understands it better), those assumptions go wrong, and the naïve storyteller ends up with a rather superficial or misdirected answer.

More often, however, briefers arrive at a wrong answer because it is the one that seems right for them. This bias can at times happen unconsciously. As the storyteller develops options, he (or she) identifies one that he falls in love with, and this emotional attachment blinds him to that option's shortcomings in a manner not unlike a parent blinded by love for a child. In more extreme cases (and perhaps more prevalent cases), the bias is conscious, and the briefer is no more than a salesman, selling the wares on hand or the solution most beneficial to the project or organization. The briefer has presumed the correct answer and attempted to build a colorful tale that supports it. Audiences are amazingly astute at seeing through this window dressing and at recognizing that something doesn't make sense. The storyteller is further challenged in this case by a lack of real belief in the story, and again, audiences have a sharp eye for a less-than-genuine performance.

Short of arriving at the wrong answer, the other common failure is to reach the correct answer via a path that is unclear or unsupported. Here the issue is that the audience is left unconvinced. The storyteller points to the end of the rope and yells "Eureka!" but the audience is looking at the rope that still remains tangled. In the worst of cases, the message itself is not clear, or even absent altogether.

The secret to the great briefing story is therefore in avoiding these two failures. The journey must be to fully understand the question or problem, avoid biases, overcome an initial lack of knowledge or information, and creatively explore options, all to arrive at the right destination—the right answer. The storyteller must understand this journey well enough to compress the voyage into an easily understood story that logically leads the audience to the same destination. But how can this tall order be achieved? The answer lies in systems analysis.

For the briefing story, systems analysis is the equivalent of novel writing. It is the process by which the themes, characters, and plot are discovered. It is part research and part innovation or creative thought. It is part rigorous modeling and mathematical analysis and part logical thinking. It provides understanding and clarity, which may then be artfully woven into a dramatic and persuasive story.

A few of you may be balking a bit at this approach to the winning story. Firstly, it requires proposing the "right" answer, and business savvy may vote against that answer if it does not offer a profitable approach or leverage your business's unique capabilities. Secondly, this approach requires the application of systems analysis, which from the short description above sounds nothing short of mystical. So, let me briefly address each of these concerns.

First, systems analysis is indeed more art than science. No formula exists to apply systems analysis, and in fact, each problem to which it is applied will likely be very unique and lead to a unique incarnation of the art. However, some approaches and best practices are common across a wide set of systems analysis applications, and knowing these demystifies the art at least a little. In addition, certain classes of questions or problems lend themselves to a more formulistic approach. One such class involves addressing a specific defined problem and seeking the recommendation of a specific solution approach; this class will be the focus of the discussion here. This class of story generation can be approached with a methodology I refer to as the *blue-team* process. The framework and best practices associated with this process provide at least a beginning template with which to conduct the required systems analysis and around which to weave the final story.

Although perfecting the art of systems analysis is not an overnight endeavor, some comfort can also be found in the fact that a little systems analysis can go a long way. The key is getting the story right in the beginning and charting the right course. That path setting is often small in scope, in terms of both time and people, in comparison to the follow-through execution of the plan. In fact, a small group highly trained in this art of systems analysis, and performing this course-charting function, can be a significant multiplier for a large enterprise. This elite group is not dissimilar in some regards to the business-development units that many organizations create, except that it acts in a very particular way, namely, it is guided by the blue-team process described below. Key in following that process is the ability of the blue team to maintain enough independence from the organization so as to arrive at answers in an unbiased manner.

This last concern brings us, of course, to the other question—what if the right answer is not what best aligns with the interests of the business enterprise? Here I have two comments. First, while it is great to bet on the winning pony, it is at least second best to know that a particular pony will not win and to not bet on it. Business strategy involves not only maneuvering to make winning bets, but also understanding what problems and markets are best avoided because they do not match well with the enterprise's resources and capabilities. Hence, the systems analysis methodology proffered here can help guide both tactical and strategic decisions by better identifying winners in advance and allowing an organization to understand how aligned or misaligned it might be with the winning approach.

Second, when done consistently and correctly, this systems analysis approach can help an organization get ahead of the crowd. Many times, the clear insight provided allows understanding of where a problem area is headed and hence what resources, capabilities, or approaches will be needed in the future. These insights are often reached well in advance of other competitors who are not benefiting from such a clear and unbiased process. Over time, the issue of being mispositioned may arise less, and an organization may find itself waiting with the right answers when the problems emerge.

The objective of this chapter is to describe in more detail some of this mystical art of systems analysis and to arm the reader with some of the tools necessary to attempt this path to not only a winning story, but a great story. Of course, in systems analysis, like many things, nothing can replace experience. Reading a good book on golfing does not make one a pro golfer, but can perhaps drop a point or two from one's handicap, particularly if one is new to the game of golf. The goal here is much the same. To further increase early improvement, the discussion here will also focus around the blue-team methodology, which for some problems will simplify the overall systems analysis approach by providing more guidance on its application.

7.3 The Power of Blue Teaming

One Tuesday afternoon years ago, while sitting in my office and wading through the usual volumes of email, I received a call from a colleague. She had been contacted by an executive of Organization X. This large entity was facing an emergency, one which I will describe here merely as a "needle in a haystack" problem. There was a particular needle that they needed to be able to identify amongst a large, diverse background (the haystack). Organization X was apparently pleading for our help and needed a solution in about a week. When I finished with a good laugh over the proposed timeline and realized that my associate was really serious, I indicated I would try to help. I knew very little about the needle involved or how to find it, but the problem sounded interesting. I am always up for an interesting problem, so "Team Blue" volunteered to dive into the crisis.

What I quickly discovered was that Team Blue was not the only organization Organization X had contacted to ask for help. Two other teams were involved, which I will refer to here as Acme Needle and Needles'R'Us. As you might guess, both of these competitors knew a lot about needles and were known to be expert needle finders. In fact, at least one of the organizations had been working on this specific problem for some time.

Despite this experience, the two other teams had very different views. Needles'R'Us, the one that had been working the problem, felt that it was nearly impossible. They had analysis that showed, in fact, how hard it was, and on the basis of this analysis were recommending Organization X approach the problem with more "brute force," which would potentially be very costly to implement and have a host of additional issues associated with it.

Acme Needle, in contrast, believed that the problem was easy. They were proposing to solve not only this particular problem but also a set of more difficult problems. They offered to show how easy this problem was and organized a demonstration to which they invited both Needles'R'Us and Team Blue. This demonstration was a disaster. Acme Needle failed to show any capability to detect needles. But importantly, it was an opportunity for Team Blue to gather some insights into the problem and return fueled to dive into the blue-team process.

A short time later—about a week (perhaps two)—Team Blue was briefing Organization X with the right answer. This winning briefing not only sorted out the disparity between the Acme Needle and Needles'R'Us positions but also provided a solution that would work. Team Blue had overcome its initial lack of experience and knowledge of the problem, and had quickly emerged as the leader in the pack. Acme Needle and Needles'R'Us had essentially dropped out of the race. Team Blue subsequently got the opportunity to implement and test out the solution they had offered, and it was Team Blue that ended as the heroes in this story, having successfully come to Organization X's aid.

Let us talk at this point a little about the process Team Blue used to come out with the winning solution in this problem. The process, appropriately named, is the blue-team process. The steps in that process are depicted in Figure 7.1, and an examination of those steps reveals a little about the general flow to the process of developing a solution. The process can be split into two major pieces—understanding the problem and innovating the solution. The first of these two pieces involves clarifying the details of the problem to be solved and understanding all the potential constraints on a solution.

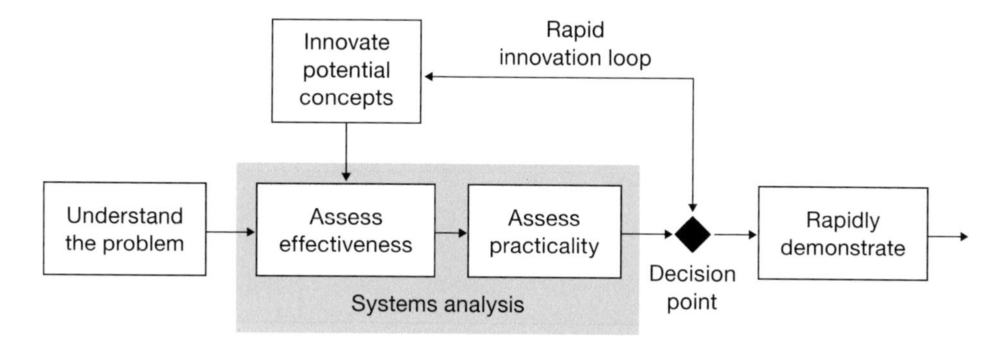

Figure 7.1 The blue-team process.

The second piece is then innovating potential options and reviewing them in terms of their feasibility—both the ability to realize them from a physical or engineering perspective and their usability in the hands of the original party who had the problem to be solved.

At first glance, these steps may not appear much different from what many would expect in solving any problem. What is so unique to this blue-team process? As is usually the case, the devil is in the details. What separates a good job from a mediocre or inadequate job can be subtle. It is the collective integration of good jobs in each step that makes the blue-team process so powerful.

Also, it may appear that the key part of the blue-team process is the second piece in which the solution emerges. It is true that this is the more creative piece. Most of us who are at all adept at problem solving love to brainstorm and to exercise our creative muscles. But long before we get the opportunity to innovate new solutions, the process may go awry in the problem-formulation stage. Understanding the problem well not only points us in the right direction but also provides key details that are fodder for innovative thought. For these reasons, we will focus the remainder of this chapter on having a good start—the problem of understanding the problem as the very first step toward the answer and powerful story we seek.

7.4 Understanding the Problem

This first step in the process would seem straightforward, but in many questions it is one of the most challenging steps and fraught with opportunities to fail early and derail the entire blue-team process. In this stage, of course, the question being asked is communicated to the blue team and clarified. Background research is often required to better understand the context of the problem, such as what has been done or considered previously, or even the definition of success in solving the problem. Generally, this step consists of asking a lot of questions, hopefully to those who best understand the problem and who will ultimately use the solution that is developed. This step is therefore similar in many ways to the surveys that marketing units do to determine what needs might be addressed by a new product offering and what the options for that product might be. That typical marketing process of understanding the problem is, however, plagued with a high failure rate. Christensen[1] points out that "Thirty thousand new consumer products are launched each year. But over 90% of them fail—and that's after marketing professionals have spent massive amounts of money trying to understand what their customers want." He goes on to postulate that this happens because marketers don't correctly understand the "jobs" consumers are purchasing products to perform. For example, in one now well-known case, he describes a fast food restaurant's effort to improve sales of its milkshakes. After extensively surveying customers on what improvements to the milkshake they might like to see (e.g., thicker, more chocolaty, etc.), and implementing some of these changes, the restaurant saw no increase in milkshake sales. Much more insight was obtained when customers were instead asked for what job they were purchasing a milkshake. It was discovered that 40% of milkshakes were purchased early in the morning, and most of these by commuters who faced long, boring commutes and who also recognized that they were often hungry again mid-morning. The milkshake filled this job quite well. It was thick enough that it took a great deal of time to drink through the small straw, providing an interesting distraction during a commute. It could be consumed with one hand, making it compatible with driving. Finally, it was filling and eliminated the mid-morning hunger. Understanding the job to be done allowed more beneficial milkshake improvements to be considered, such as a self-serve machine to speed purchases for rushed commuters or the addition of small pieces of fruit to further add interest to the milkshake during the otherwise boring commute.

As in this milkshake example, not understanding the job when implementing the blue-team process can lead to launching the subsequent creative innovation steps in the wrong direction. In some cases, the assumed problem is simply wrong; in others, it is either much easier or much more challenging than the actual problem. For example, in one type of failure that I refer to as the sandbox failure, the problem is misinterpreted as a simpler, unrealistic problem; the actual problem is more difficult in one or more ways. This failure typically happens when there is insufficient knowledge of

1 C.M. Christensen, S. Cook, and T. Hall, "Purpose Brands and Disruptive Innovations," *Harvard Business Review*, vol. 83, no. 12, 2005.

the issue, and the analyst therefore makes up a great deal in its definition, arriving at a creation that is often simpler than reality and that leaves out many of the real constraints. Illustrative of this failure, a colleague of mine working for the military once shared with me his work on automatically recognizing military targets from images. He showed me images of various military vehicles sitting in a large grassy field and went through example after example of his technology's ability to identify the vehicle. After complementing him on his technology, I felt compelled to ask how often real military vehicles were found in the middle of grassy fields. Not surprisingly, the answer is, not very often; the more realistic case involves vehicles hidden through a variety of means. Also not surprisingly, the technology that I had been shown did not work well at all for those realistic cases. The problem of "How can military vehicles be identified?" had been the stated goal, but a sandbox version of the problem had been adopted, and the realities of what make that problem challenging were not recognized (or were ignored).

In some instances, of course, a simpler sandbox problem can be adopted deliberately as a first step when the actual problem appears overly challenging to take on immediately. Although this approach is useful at times, care has to be taken that one does not avoid the real problem or that one does not head down a path that must be abandoned with a restart when the real problem is addressed. To be most useful, the simpler first step must begin to take on some of the issues that make the real problem seem intractable, or else little actual progress will be made.

The opposite of the sandbox problem is the "hot snow load" failure, in which the problem is envisioned as very difficult when, in fact, the actual problem is not nearly as bad. Often this failure happens when, again, the analyst has too little information and imagines all the possible cases that might arise in the described problem. The analyst then sets out to find a general solution to all of these possible cases even though many are unimportant to the eventual user or will happen so infrequently that their consideration is not warranted. For example, if you were to consider the problem of flat tires, you might worry not only about a single flat but also about the possibility of a second flat tire occurring while on the way to get the first repaired, or similarly, the possibility of receiving two flat tires in the same event. You could even begin to justify these concerns by considering realistic scenarios in which they might occur. Despite these possibilities, however, most drivers would not consider trading cargo capacity to carry two spare tires or replacing the spare tires with a do-it-yourself patch kit that allowed handling a virtually unlimited number of flats. They would define the problem as one of needing a contingency for a single flat. Sorting out which cases matter and which don't is easy in an example of this kind; you simply have to ask drivers and they will tell you. The failure of over-scoping the problem only occurs when the analyst fails to ask.

What is harder is the case in which the user is the one communicating a desire to handle all of the cases, including ones that seem very unlikely. It is possible the user understands the problem well and has thought through the need for all these cases. Some of them, while low in likelihood, might be significant in cost should they occur. For example, in the flat tire example above, the need to address the more unlikely cases might change if the vehicle is to be used in the remote wilderness removed from communication with sources of help. In that application, planning for multiple flats might be a good idea and a real requirement in the mind of the user. Many times, however, a user asking for many irrelevant cases is an indicator of a user who doesn't understand his or her own problem well. In the absence of understanding which cases are important, the user is asking for all of them. The problem here is harder because the analyst now needs to take on the job of defining the problem not only for him- or herself, but for the eventual user. Unless there is an easy solution that handles all of the cases, the analyst must sort out which cases are important and address only those cases in subsequent innovation steps, lest the challenge of innovating a solution become intractable. The analyst must be able to convincingly justify to the user this narrowing of the problem scope, or else the solution is not likely to be accepted. The real challenge in this sorting of cases is not in examples for which it is obvious that more or less irrelevant cases are being requested, but in problems for which it is not so obvious which cases are important. It is often necessary to begin the innovation process, discover that the broad scoping of the problem is proving too challenging for a reasonable solution, and then revisit the problem-definition step to examine whether the breadth is necessary or if a solution of a narrower problem would provide acceptable (or even significant but partial) capability. Hence, many times, understanding what the real problem actually is and clearly articulating the reasoning for that definition to a user who doesn't fully understand the problem represents a very important part of the systems analysis conducted to execute the blue-team process.

Another important case leading to an overly challenging problem definition occurs when the user understands the problem but chooses to levy additional, less critical requirements because of not understanding the cost these requirements bring or because of lusting after the ultimate gold-plated solution. In the former case, the user is simply naïve to the challenge of meeting all the nice-to-have but noncritical requirements, and adopts a "Why not ask for everything?" attitude. In the latter case, the user is akin to the car buyer who looks at increasingly luxurious or sporty models and returns from the showroom with the top of the line. The heated leather seats and sport trim probably are not needed for running errands around town or picking the kids up from school, but the buyer has fallen in love with the fancy model, and all of the unnecessary options become irrational requirements. Buying gold plate is fine if the buyer can afford it, but the rest of us need to decide what we really need, what the real requirement is. Voltaire said it well: "*Le mieux est l'ennemi du bien*," or "The best is the enemy of good." In many problems, it falls to the analyst to explain to the user why there is insufficient technology or funding to have "the best." The 80% solution is often good enough and obtainable with far less than 80% of the cost, effort, risk, and technology required in the 100% answer. The analyst's dilemma is determining what constitutes 80%, and "selling" the user on the idea of accepting something less than the luxury model.

A final type of failure worth mentioning is what I refer to as the bland problem definition. As in many of the above cases, this failure arises from a lack of detailed information or understanding of the actual problem, but that shortfall manifests itself here in the form of a vague problem description. The impact of this failure on subsequent innovation steps is to effectively broaden the problem definition and to lead to some of the same difficulties described above. But there is another challenge that arises with a bland or vague problem description; the lack of detail provides little fuel to start the innovative thinking process. Innovation often happens when some small detail of the problem is exploited, and in the bland problem description, there are no such knobs to grab hold of and exploit. For example, consider the difference between "solving hunger" and "solving hunger in a specific place affected by an unexpected summer drought." In the latter case, you are probably already thinking, "How can additional irrigation resources be applied?" or "Can shipments of food be arranged to get through the summer until precipitation increases in the winter months?" In contrast, in the former problem, you are probably thinking, "Where do I even start?" In fact, that exact thought—not knowing where to start in innovating a solution—is a pretty good indicator that you have a blandly defined problem and need to dig more for the details that better describe the real problem. Of course, someone could really ask you to think about the problem of world hunger, but even then you would want to understand a lot more about the issue (e.g., what causes it, how many are affected, what approaches have already been tried to reduce it, etc.) before ever starting to innovate new solution ideas.

A common theme in many of the failures above is a shortage in information, detail, or understanding surrounding the problem. It would seem simple enough to avoid these shortages by asking more questions. In reality, that interaction with the eventual user is rarely so straightforward. For example, in most business environments, it is the marketing department that interacts with the customer and asks the questions, and the team developing new products or solutions hears the customer's thoughts as translated by the marketing department. The one or more interpreter stages between the user and the analyst can lead to significant misrepresentation of the actual problem. As illustration, consider the simple children's game in which several children line up in a row and an adult whispers a single-sentence message in the ear of the first child. The first child in turn whispers the message to the second, and so on, until the end of the line is reached. The final child then repeats out loud what he or she believes the message to be. Almost invariably, the pronounced message is a highly garbled (and amusing) version of the original message. This distortion of the message can be achieved with very few children in line. While adults may perform better than children in this game, it remains that it is difficult to communicate facts when they are not understood, and in many instances, the interpreter positioned between the user and analyst does not understand the problem well—certainly not as well as the analyst would want to before launching an innovation of a solution. Even if the interpreter successfully parrots the problem as described by the user, he or she is in a poor position to answer further questions the analyst may ask. Hence, there is a risk that the interpreter will make up an ill-informed depiction of the problem (failing in ways similar to those described above for the analyst), and then pass that faulty description along to the analyst. The key to avoiding these problems is to place the analyst who is working to understand the problem in direct communication with the folks that have the problem. Although obvious, it is rare that this approach is actually implemented.

In addition to direct communication, it is also important that the communication be continuously maintained throughout the blue-team process. Although the diagram of Figure 7.1 shows a linear process between the first two steps (e.g., understanding the problem and innovating potential solutions), there is in practice more iteration than what is shown. As mentioned briefly above, it is frequently necessary to revisit the understanding of the problem throughout the blue-team flow, changing scope as appropriate and needed to enable solutions, or to better understand problem details that become important for particular solution implementations.

In many cases, the issue of usability further forces a revisiting of the problem definition. Particularly in highly technical problems, a tendency exists for the analyst to focus on the technical aspects of the problem solution—whether the solution could theoretically work. In those cases, the user's constraints, which define how a solution might or might not be used, and the concept of operations for that solution often receive less consideration. These issues are equally a part of the problem definition, but it can be difficult to ask all the questions regarding these aspects until a candidate solution is in hand. Hence, the analyst often gains more understanding of the problem when a solution approach is socialized with the user.

7.5 Summary

Great stories are things of power—clear and convincing, but above all, correct in the message they deliver. Arriving at that correct message is no small thing, considering the complexity, constraints, and biases that we must overcome in most problems. No mechanistic formula exists to bring us to a compelling, correct answer, but in some problems the guiding approach the blue-team process provides can at least help keep us on track. That track starts with a detailed understanding of the problem. That first step has been the focus of much of the discussion above because it is a common place to misstep early and can derail all that follows. The track ends, done right, with a great story that guides the audience to the right answer.

About the Author

Robert G. Atkins is the head of the Advanced Technology Division at MIT Lincoln Laboratory. Prior to this position, he served as the assistant head of the Intelligence, Surveillance, and Reconnaissance and Tactical Systems Division and, earlier, as assistant head of the Homeland Protection and Tactical Systems Division and as leader of the Advanced Capabilities and Systems Group.

He began working at the Laboratory as a cooperative education student and research assistant in the Air Defense Techniques Group, where his work focused on the modeling of electromagnetic scattering and radar cross-section prediction. Upon completion of his doctorate, he joined the technical staff of Lincoln Laboratory. He became assistant leader of the Systems Analysis Group in 1999 and associate leader of the Sensor Exploitation Group in 2000, before returning to the Systems Analysis Group in 2003.

Dr. Atkins has expertise in the areas of radar cross-section modeling, automatic target recognition, and the systems analysis of air defense and surface surveillance systems. In his current position, he leads efforts aimed at the development of revolutionary component and subsystem technologies to enable new system-level solutions to critical national defense challenges. He holds SB, SM, and PhD degrees in electrical engineering from the Massachusetts Institute of Technology.

8

Truth and Systems Analysis

Alan D. Bernard

> When you are studying any matter, or considering any philosophy, ask yourself only what are the facts and what is the truth that the facts bear out. Never let yourself be diverted either by what you wish to believe, or by what you think would have beneficent social effects if it were believed. But look only and solely at what are the facts.
>
> —Bertrand Russell, 1959

8.1 Introduction

My initial work at MIT Lincoln Laboratory, many years ago in 1972, was working on flight testing and modeling the environment that surrounded objects entering the Earth's atmosphere at Mach 23. This work was entirely consistent with my background in physics. When I started in systems analysis, I was astounded at how little of my extensive schooling in math and physics was needed. Though the inputs to the systems analyses often required a considerable math background, most of the actual systems analyses merely required algebra or perhaps elementary calculus. I wondered why I had an advanced degree in math to perform analysis with arithmetic, algebra, and calculus. About that time, I came across a book entitled *All I Really Need to Know I Learned in Kindergarten*. The sentiment seemed so true, and maybe there actually is some truth to it.

The problem is that systems analysis involves more than merely manipulating variables and inverting matrices; it involves thinking. Trying to get one's arms around ill-defined problems involves a multitude of skills and processes that cannot be predicted ahead of time. It involves defining the question that is of interest, that is important, and that is answerable: no mean feat. On top of that, once the results have been determined, one has to construct a story that will represent the findings in a coherent, understandable, and convincing way. At the end, one would like to hear both "It doesn't look that hard" and "Of course that is the answer." With those statements, one can declare success.

So, one might think of an advertisement for hiring a systems analyst as being something like the following: "Needed, advanced degrees in math, physics, engineering, and political science, though they will hardly be used. Will work long hours to answer questions that are both uncertain and unanswerable. Will find disagreements from sponsors, stakeholders, and one's own management structure. Will get little thanks from labors but will be faced with additional work resulting from analysis." That description doesn't look like a path for enticing top-notch talent or for advancement. However, I note that the small group I led in systems analysis has produced, so far, eight group leaders and three division heads at the Laboratory, four if you include me. It seems that the Laboratory finds systems analysis good training for management roles. A quote from Steve Grundman of the Atlantic Council indicates that this training may have been recognized as very valuable: "Compared to finding a good systems engineer, it's comparatively easy to build a factory, an insight I'm not sure was so widely shared 20 years ago as it is today."

8.2 Truth and Systems Analysis

Systems analysis as a discipline seeks to determine overarching "truths" from disparate and often ill-defined data. It has been my experience that the practitioners of systems analysis are dedicated to finding out these truths. In some organizations, systems analysis departments are closely coupled to an arm of marketing. It is well to explore just how the needs of marketing are traded off with the discipline's need for objectivity.

The systems analysis art at Lincoln Laboratory has evolved to include a number of significant characteristics. First, systems analysis should tell a story having a beginning, a middle, and an end. The beginning tells why we care and what we are telling, the middle tells what and how we have learned, and the end tells what it all means and why we should care.

Second, the analysis requires transparency. We should be able to see not only what the results are, but why they come about. One consequence of this requirement is that some analysts tend to have a healthy skepticism of large-scale computer simulations that obscure the inner workings of the analysis.

Third, models used in the analysis should be connected to verifiable tested data. When such data are unavailable, test programs should be constructed to provide the vital foundation of a valid systems analysis. This need accounts for the numerous transfers back and forth within the Laboratory of people in the analysis and test disciplines. (A number of my author colleagues emphasize this important role of testing.)

Fourth, one should strive for an analysis that is more than just a correct manipulation of facts and mathematics; the analysis should convey the essential "truth" in its findings. This demand for truth is where the difficult thought and judgment come in, and protocols for capturing the illusive truth are the main focus of my writing here. I will make frequent use of examples, shortened and unclassified to fit this context, to illustrate this search for truth.

8.3 Objectivity

> Lies, damn lies, and statistics.
>
> —Mark Twain

The question arises, where should we put systems analysis in the progression above? There is plenty of room well to the left of *lies*. One can place truth, half-truths, inadvertent lies, and even white lies to the left of lies in the progression of deception.

Every systems analyst will seek to describe his or her product in a favorable objective light; however, one needs to assess where each analyst actually falls in objectivity. That is not to say that the analysts actually lie; it has been my experience that overt lying within the systems analysis community is very rare. Of course, there are two kinds of lies: lies of commission and lies of omission. It is seldom that systems analysts commit lies of commission; it is certainly usually the case that their stories tell the truth but perhaps not the *whole* truth. For, after all, systems analysis is about storytelling: telling stories to help the reader understand complicated multidimensional problems, but often a specific agenda creeps in.

My experience in observing many systems analysts' products is that the general systems analysis is often, to the left of lies but perhaps not all the way to the left at truth. We will examine some cases that illustrate this stretching of the truth and some things we have observed that can help pull the story toward the truth.

Often, the appearance of lies or even damn lies is not a result of lying per se, but of the lack of diligence on the part of the systems analyst—perhaps answering the question asked rather than answering the question that should have been asked, or making assumptions that tilt the results in a way that is not intended, or even illustrating the results in such a manner that one is misled as to the true import of the conclusions. These shortcomings nearly always suffice as reasons for errors in results rather than the starker prospect of a lie, using the definition that to lie is to state something with disregard to the truth with the *intention* that people will accept the statement as true.

In the following pages, we will review some of the issues arising in systems analysis that keep the stories from making it all the way to the domain of truth. The three areas I will describe are the question that is asked, the assumptions that are made, and the display or presentation that tells the story.

8.4 The Question

> To find the exact answer, one must first ask the exact question.
>
> —S. Tobin Webster

The story told by a systems analyst is usually in response to a question, or has within itself the implied question that the story seeks to answer. It is vital to have a question that is truly what one wants to know and that can reasonably be answered by a detailed systems analysis. We can illustrate the importance of the question with a couple of examples that have occurred in evaluating weapon systems.

8.4.1 Fighter Comparison

A question of importance to the U.S. Air Force was how a foreign fighter (we'll call it the Red fighter) compared with the current U.S. air superiority fighter (the Blue fighter). In the analysis performed, the Blue fighter did not fair too well. The analyst reviewed a number of fighter attributes and evaluated each fighter for that characteristic:

- Top speed: important for engaging in and disengaging from combat
- Maximum radar display range: important for earliest detection
- Weapon load: important for multiple shot opportunities
- Maximum altitude: important for capturing the high ground to "shoot downhill"
- Turn rate: important for maneuvering to avoid missiles and to enter a firing geometry

The Red fighter had a higher top speed, and its radar display could show targets at longer ranges than the Blue fighter. The Red fighter carried more weapons (it was a larger aircraft) and could maintain level flight at a higher altitude. The Blue fighter did have a higher sustained turn rate than the Red fighter. So, we can see that the Red fighter "won" 80% of the above parameters, all but the last one—turn rate—and was judged to be the better fighter. So, what can we say in answer to the question, "Which fighter is better at these five characteristics?" We can say, honestly, the Red one.

An analysis of air-to-air combat might tell a different story. It turns out that top speed is not a good indicator of fighter capabilities. Combat between fighters seldom occurs at their top speed, with afterburners consuming prodigious amounts of fuel, but rather at transonic speeds and mid-altitudes. The maximum radar display range does not indicate the maximum detection range, a far more important measure. The engagement range of the weapons may be far more important than merely the number carried. The analyst might then select five other indicators:

- Maximum transonic acceleration: important for close combat
- Maximum radar detection range: important for earliest detection
- Maximum weapon engagement range: important for first kill
- Maximum weapon load: important for multiple shot opportunities
- Maximum turn rate: important for close combat

By this measure the Blue fighter won 80% of the above parameters, all but maximum weapon load, and would be judged the better fighter.

If we change the question to "Which fighter is better *at combat-related characteristics?*" we would, of course, use the second list as representative of combat-related characteristics. Thus, by a small change in the question, we can give an answer that is true to the (believed) intent of the questioner. Furthermore, the answer is closer to the "truth" that, in fact, the Blue aircraft was superior to the Red. Subsequent combat experience between these two aircraft reinforced that belief that the Blue fighter had better combat-related characteristics, as witnessed by its superiority in actual combat-related exchange rate.

8.4.2 Advanced Aircraft Loss Rate

When stealth was a new idea, Lincoln Laboratory was asked to calculate the expected loss rate of a stealth aircraft against a future air defense system. This analysis involved a vast number of modeling exercises. The advanced threat had to be postulated since it was to represent the best that could be built some 20 years in the future. The defense's terrain was modeled so that the interaction of the radar signal with the ground, producing clutter to the radars and modifying the propagation of the radar paths through space, could be determined. The future surface-to-air missile (SAM) system's capability was modeled as was the capabilities of advanced fighters. The results of all this modeling were then put together to determine the expected loss rate of the stealth aircraft.

The result of this analysis was a series of graphs illustrating the aircraft loss rate under various threat lay downs and missions. The surprising result was that the advanced aircraft would suffer considerable losses in accomplishing its mission against the future advanced threat. We checked the mathematics and could find no errors to explain the large-scale attrition to be expected from such engagements.

Of course, harkening back to the point of the previous discussion, we wondered if this was *really* the truth. Fortunately, our modeling was transparent in that all the effects modeled could easily be viewed and checked to see their influence on the overall conclusions.

To test our analysis, we looked at both advanced and legacy aircraft performance actually achieved in recent combat against contemporary defenses. We found that we significantly overestimated the performance of defenses against the models of the actual legacy aircraft flown in combat! The advanced aircraft suffered very little attrition against these modeled actual defenses; however, because the models did not adequately give the known attrition of legacy aircraft, they were deemed unreliable in giving the attrition against either contemporary or future defenses. All we could see was that the advanced aircraft was much more survivable than the legacy aircraft against any postulated defenses.

Defensive systems require a continuous series of successes at each stage of the so-called defense kill chain. Any imperfection in any stage of the defenses will diminish the effectiveness of defenses. Neglecting or minimizing any of these imperfections in a model will bias the modeled results toward exaggerating the defense's capability. Since all of the imperfections of a defense are not generally known (even to the defenders), we generally exaggerate the capabilities of defenses in analysis. This effect, often couched as the fog of war, explained our inability to predict the past combat history.

One example we encountered was the discrepancy between the advertised values of SAMs' effectiveness and combat results. SAM systems are often advertised as having a given single-shot probability of kill (P_k), a number derived from flight-testing the systems. This number might vary from system to system, but is generally quoted as larger than 0.5 and often as high as 0.9. If a system is at the lower end of things, then firing two missiles seeks to compensate for such a shortcoming. Two 0.5 P_k missiles (assuming independent errors[1]) will give a combined P_k of 0.75, two 0.7 missiles yields 0.9, and two 0.9 missiles gives a 0.99 probability of kill! So one can migrate to quite high probabilities of kill by a simple selection of the assumed value of P_k.

The facts of combat paint a different picture. During war in the Balkans, the Serbians fired 815 SAMs at coalition forces. With the above P_ks, we would expect between 407 ($P_k = 0.5$) and 734 ($P_k = 0.9$) coalition aircraft to have been shot down; in fact, it was but two, indicating an overall P_k of 0.002! This is a far cry from the advertised values.

To add to an analyst's complexity, values of P_k are not static because of changes in tactics and equipment. For example, during the Vietnam War, we saw a dramatic change in P_k as the United States learned to operate against surface-to-air missile systems. During Operation Rolling Thunder, the missile P_k was 0.092. Later in Linebacker I, the P_k was 0.016, and in Linebacker II it was 0.014.

We finally recognized our inability to account for every nuance that might diminish the capability of a defensive system and determined we could not honestly answer the question as posed. We resorted to answering a new question: "How does the postulated advanced aircraft's survivability compare to that of legacy aircraft?" This question was answerable by carefully modeling each aircraft, optimizing its tactics for its particular capabilities. The supposition is that, though the actual calculated loss rates are probably incorrect, the relative performance will tell which is *relatively* better. When we used this approach, we concluded that the advanced aircraft was far more survivable than the legacy aircraft.

8.5 The Assumptions One Makes

> Euclid taught me that without assumptions there is no proof. Therefore, in any argument, examine the assumptions.
>
> —E. T. Bell

A friend proficient in financial matters once told me that to read a profit and loss statement, one should first read the footnotes. All the essential truths lie there. In the same sense, to understand any systems analysis, one should first look at the assumptions. We often joke, "A systems analyst will assume everything but responsibility." We, of course, know that is not true. Nonetheless, as with all aphorisms, there is a

1 The presumption of Gaussian distributions and of independence, though often erroneous, is so common as to generally not even make it to the footnotes. Here, we do, for completeness, caution our readers of these, often important, mistakes.

germ of truth in the statement. The following example illustrates the importance of examining assumptions.

8.5.1 Radar Coverage Example

To evaluate the capability of radar, it is convenient to display a radar coverage diagram that merely shows where a radar has a particular performance level, displayed as a curve on a plot of altitude against ground range. An example of such a graph is shown in Figure 8.1. We see that an aircraft flying at an altitude of 10,000 feet is detected at 180 miles, though one at 40,000 feet is detected at 170 miles. We notice too that an aircraft flying at 60,000 feet is not detected at all. This plot leads us to the proposition that high-altitude flight would be very valuable if it could result in overflying the defenses (in this case the defenses' surveillance capability). The assumption that high-altitude flight could be undetected was one reason for the early deployment of the U-2 aircraft.

Before we insist on the large wings needed for very-high-altitude flight, let us examine our assumptions. Hidden in Figure 8.1 is a note that the curve represents P_{DSS}, or single-scan probability of detection, of 0.5. That value indicates that inside the curve the P_{DSS} is higher than 0.5, and outside, lower, but *not necessarily zero*!

Figure 8.2 illustrates, in addition to the P_{DSS} curve of 0.5, those curves for 0.9 and 0.1. If we examine other values of P_{DSS}, as in Figure 8.2, we see that the 60,000-foot aircraft at 180 miles has only a P_{DSS} of about 0.3, (about half way between the curve for 0.5 and that for 0.1), as opposed to 0.5 at 40,000 feet. That is a lower detection probability, but it certainly is not zero.

A more useful metric might be the cumulative probability of detection that represents the probability that an air vehicle will be detected before a given range. The cumulative probability of detection for aircraft flying at 500 miles per hour is

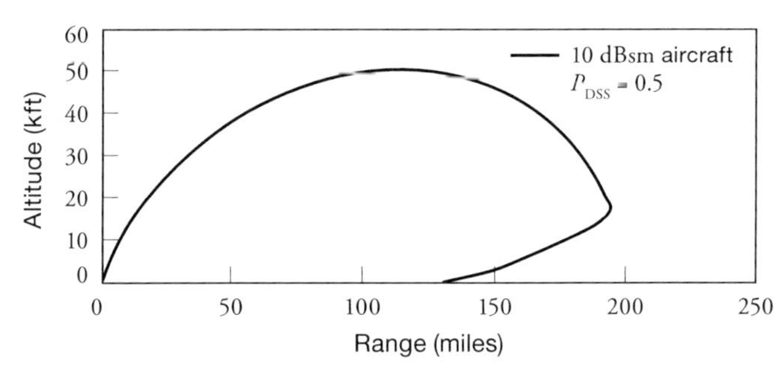

Figure 8.1 Altitude versus ground range.

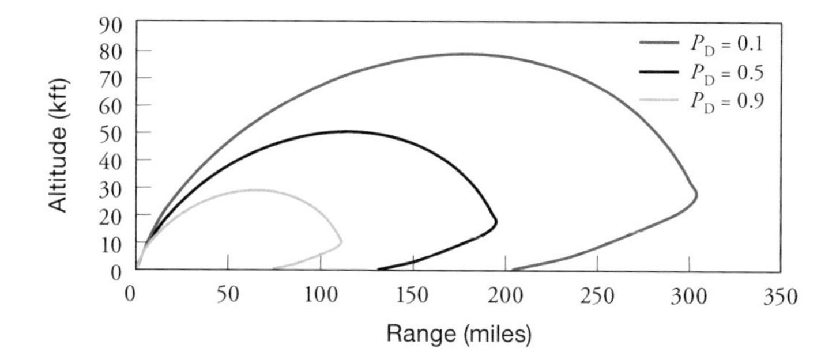

Figure 8.2 Multiple detection probabilities, P_D.

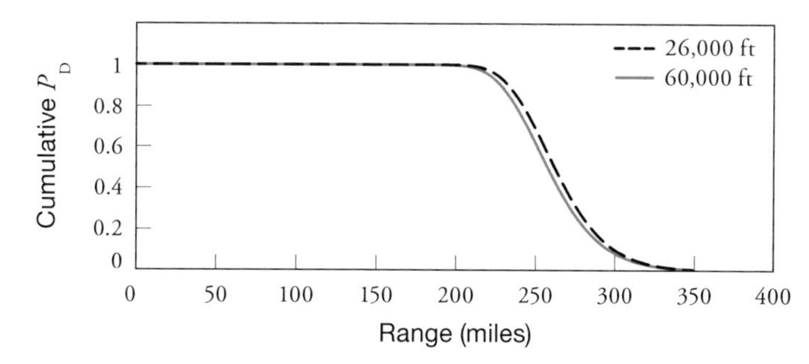

Figure 8.3 Cumulative detection probabilities.

illustrated in Figure 8.3. Examining the cumulative probability of detection, we see that there is not really a lot of difference between altitudes of 26,000 and 60,000 feet, and our *assumption* that the single-scan metric P_{DSS} of 0.5 describes the radar coverage was a particularly poor assumption.

A further assumption that has been made is the radar cross section (RCS) of the aircraft that the radar is detecting. If we change it from the 10.0 square meters *assumed* in Figure 8.1, to 1.0 square meters (perhaps a cruise missile instead of a manned aircraft), we would have a radar performance of 80 miles as opposed to our early estimate of 180 miles!

So we see that by changing our assumptions, we change the outcome of the graphs. What we need to do is determine what assumptions tell the true story, and that is no mean feat! As we saw earlier, we must select the assumptions that tell the true story for the question that should be asked, and the viewpoint of the questioner is important. For example, if the above radar is being evaluated for an early-warning mission, the 90% probability of detection range of an adversary's air vehicle might be appropriate. If the question concerns an adversary's radar's ability to detect our aircraft, the detection range for a 10% probability of detection might be appropriate. These two numbers reflect the different outlook between the defender (wanting a high probability of defense) and the attacker (wanting a high probability of penetrating the defenses).

8.6 Consider the Adversary's Reaction

> For every action, there is an equal and opposite reaction.
>
> —Sir Isaac Newton

Many times we end up with measures of effectiveness of our current or future forces that can be used to choose among various options or size forces for particular goals. Often neglected is that the adversary would be expected to react to our various options in different ways, perhaps changing the calculus of the various results. The reaction may be indirect and may be disproportionate. An example of this situation occurs in ballistic missile defense (BMD) when we see that an adversary has several ways of attacking, each having a different outcome, and the defense can respond in a different manner to also change the outcome.

A particular deployment of BMD assets is often valued by the "cost" it imposes on the offensive forces. At some point, offensive forces can be so large as to saturate or overwhelm defenses; however, the BMD system may make that unaffordable. The cost imposed on the offense by a BMD system might be expressed as the number of extra attacking missiles needed to arrive at the same outcome as if there were no ballistic missile defense. In the next section, we will explore how this cost might change substantially depending on the adversary's reactions to the deployments.

8.6.1 Baseline Missile Defense Example

To illustrate the importance of action and reaction, we will consider a particular role for a BMD system, defending a number of our own ballistic missiles (e.g., housed in "hard" silos) to deter an adversary. We will assume that we have 10 ballistic missiles being defended by a BMD system. In a simple, ideal case, it would take the attacker 10 adversary ballistic missiles to destroy our cache of ballistic missiles if there were no BMD system (see Figure 8.4). If the defender wishes to change that, the defender might employ a BMD system to cause an attacker to allocate more missiles so that an attack becomes prohibitively expensive and therefore not a reasonable option.

A primitive defense might consist of 10 interceptor missiles situated in two interceptor farms and two associated radars to guide the defending interceptors. Now, the defense can allocate a single interceptor to each of the missiles attacking each

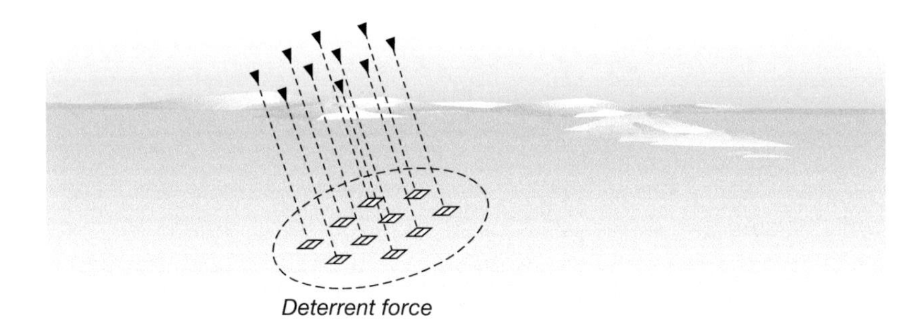

Figure 8.4 Attack on the ballistic missiles.

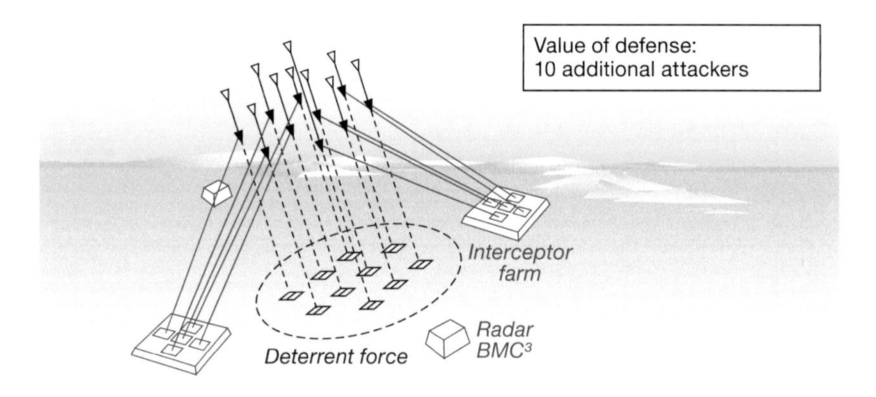

Figure 8.5 Ballistic missile defense. BMC^3 is battle management, command, control, and communications.

ballistic missile, so that (again, with perfect systems) after the attack, all the missiles survive, and the attacker has achieved nothing but to extremely annoy the defender for launching 10 attacking missiles.

In this case, the attacker, realizing the situation diagramed in Figure 8.5, could react by doubling the attack to 20 attacking missiles, so that after the defense described above, 10 missiles are left to attack and destroy all the defender's missiles, achieving the attacker's goal of preventing retaliation. Under this case, the BMD defense could be said to cost the attacker 10 missiles (20 needed minus 10 needed without defense).

8.6.2 Preferential Defense

The above case assumed the defender did not react tactically to the attack. However, the defense need not defend uniformly; it could preferentially defend half or even a single missile, to assure that something would be left to retaliate. The defender could defend half of its missiles, each with two interceptors. Under this case, the attacker would have to attack each of the defended missiles with three missiles (two for the interceptors and one left over to attack the missiles). Since which half would be defended is unknown to the attackers, all would similarly require three attackers, for a total of 30 attackers. Now we see the value of the defense is 20 missiles—20 for the potential interceptors and 10 for the now undefended missiles. This scenario is illustrated in Figure 8.6.

In the above case, we considered defending half of the missiles. We could defend any number we choose, perhaps only defend one, in which case the value of the defense is now 100 missiles! If the defense defends only one missile, 11 attackers will be required against each missile, resulting in a value of 100 attacking missiles.

8.6.3 Farm Attack

The reaction of the attacker to this cost might be to give up and live with neighbors in peace and tranquility (if not with love). Alternatively, the attacker might seek to react in a different manner and perhaps attack the interceptor farm, forcing the defenders to *waste* interceptors on the attackers. The defense must engage the missiles attacking

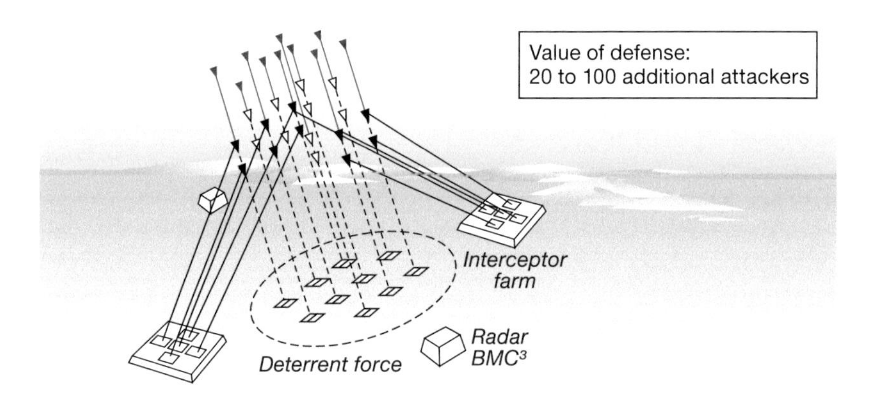

Figure 8.6 Defenders react with preferential defense.

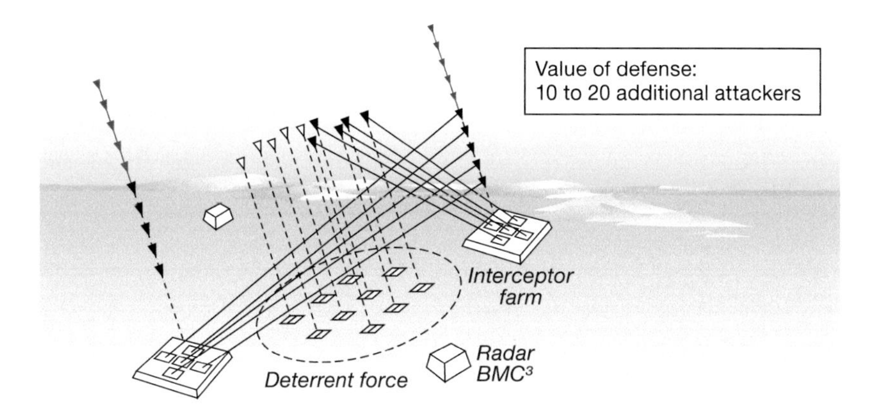

Figure 8.7 Farm attack followed by switch to silos.

the interceptor farm, or the defensive missiles will be destroyed on the ground, and 10 attacking missiles against the interceptor farms would exhaust the interceptors (either by intercepting the attackers or waiting to be destroyed). The defender would have to use all 10 of its interceptors to defend the farm, with none left to defend the missiles it was there to defend. Then, the 10 attackers would strike against the missiles, for a total of 20 attackers. The resulting BMD value is now back to 10 (the 20 attackers needed minus the 10 needed in an undefended case).

But the reaction of the defender to this could be to only defend one of the interceptor farms with all the interceptors. Now the attacker would have to attack each farm with 10 attacker missiles and then the 10 defended missiles with one, for a total force level of 30 attackers. Now the cost of the BMD system is 20 attacker missiles. This scenario is illustrated in Figure 8.7.

8.6.4 Radar Attack

Similarly, the attacker could attack the two radar sites with the same results as the farm attack. The attacker's concern would be that though both radars would be destroyed by an attack of 20 missiles against the radars, the defender's radars might have enough range such that radars from adjacent complexes might be used to guide the interceptors held back from the fight, sacrificing their radars. The radar attack would, of course, end with a BMD system value of 20!

8.6.5 Summary

We have seen a simple case of a BMD deployment that, absent any reactions by the attackers or defenders, was valued at 10 attacking missiles. Varying the tactics varied the value at 10, 20, 100, again 10, or 20 attacking missiles. In fact, the 20 value was the only one enforceable regardless of what either side chose to do. So each defending interceptor cost the attacker two attacking missiles. It is apparent, then, that to truly determine the value of a particular strategy (in this case BMD), we must keep in mind the logical allowable responses an adversary might make and what our responses would be to their tactics.

8.7 Not Just Correct But True

> "The question is," said Alice, "whether you can make words mean so many different things."
>
> —Lewis Carroll, *Through the Looking Glass*

Affirming to "tell the truth and the *whole* truth" is not only an easy option, but a generally accepted practice in any systems analysis. Perhaps truth, like beauty, is in the eye of the beholder. One person's truth may not be another's. There is a requirement in the oath of courtrooms to not only tell the truth but to tell the whole truth and nothing but the truth—a much more severe burden. This is a burden the systems analysis community must carry as an integral part of their analysis. In the following section, we shall explore some examples of the need for not only the truth but the whole truth.

8.7.1 Understatements

Figure 8.8 illustrates an evaluation of a system that, in the ideal, meets 95% of its requirements. However, as designed, it can only accomplish 82%, and as built, only a mere 42% of its requirements. The "bumper sticker" tells the bottom line: "The system does not meet all its requirements." Assuming the requirements are valid and that the analysis was conducted correctly (i.e., the arithmetic is correct), then the bumper sticker is a true statement. However, it is a statement that, perhaps incorrectly, leads one to be concerned that there is a problem in that all the requirements are not being met. Let us explore that further.

So, what is the *whole* truth? What if the bumper sticker had said, "The system does not even meet 50% of its requirements," or if it had said, "The achieved requirements show large problems in building the system," would that be true? What if the truth were, "This system meets all of its *important* objectives, but merely 42% of its total requirements"? One can ask which of these true statements captures the whole truth, that is, which is the truer statement? We might call the message in Figure 8.8 merely an *understatement* of the truth, but we shall see that though this is true, the statements—and in fact, the values in the chart—are misleading.

Firstly, as we know, all requirements are not created equally. A requirement to "survive in combat" cannot be equated to a requirement to "post all purchase requisitions online." We were taught (in kindergarten?) that we cannot add apples and oranges (or, one would think, dissimilar requirements), but nonetheless we do, in this chart, add dissimilar requirements. Achieving 42% of the requirements begs the question, which requirements? Were they the ones to care about, or not?

Furthermore, it is possible that some requirements were almost achieved while others are not even approximately met, for instance, perhaps 10 feet short of required altitude, but 500 miles short of required unrefueled flight range. In examining the achievement of requirements, we must consider the validity and worth of each requirement as well as the degree to which each is met, and then report the weighted consensus as to the overall achievement of the requirements. This accumulated requirements chart does not do that.

Maybe all we can say from Figure 8.8 is that not all the requirements are met, as we expected. The *as we expected* is added to ensure that we alert the reader to not draw the conclusion from this chart that we have (or do not have) a problem with the requirements. In fact, this chart and charts like it should be banned because, though they may be true, they can easily lead one to the erroneous conclusion that there is a problem with the requirements. There may be a problem with the requirements, but just reporting percentages, as in this chart, does not make that implication.

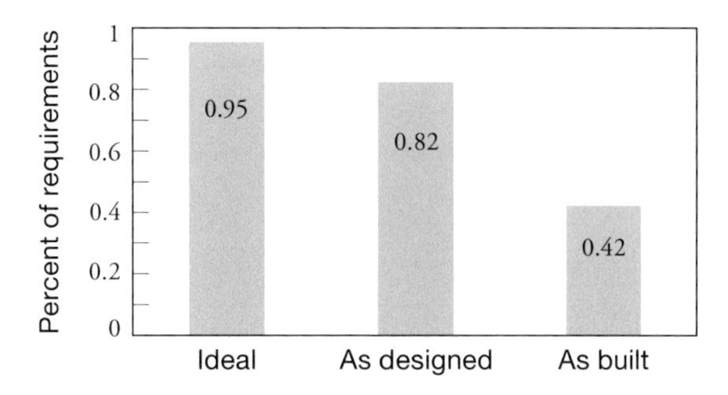

Figure 8.8 System requirements.

8.7.2 Broken Scales

The lie of the broken scale is, of course, not really a lie. One might argue that it is only misleading if one were to carelessly read a chart. I submit, though, that it is common (way too common in the news media) to show results to deceive rather than to inform. Figure 8.9 illustrates an example of a broken scale. We see a large jump in an aircraft's total predicted cost occurring in 2007. In that year, we might add a bumper sticker that says, "Huge increase in cost occurring in 2007." Of course, *huge* is, at best, an ill-defined term, and what one person says is huge may not be another's opinion. Nonetheless, we all speak the same language, and in imparting true information, we depend on a generally consistent meaning of words, rather than an arbitrary definition.

Perhaps, as we look at Figure 8.10, the same data as shown in Figure 8.9, we might come to a different conclusion, specifically that there is not a significant change in the total aircraft's cost occurring in 2007! And, there is certainly no huge increase in predicted cost. This conclusion is in opposition to the earlier statement. Furthermore, in this case, there is no question as to which is the correct statement.

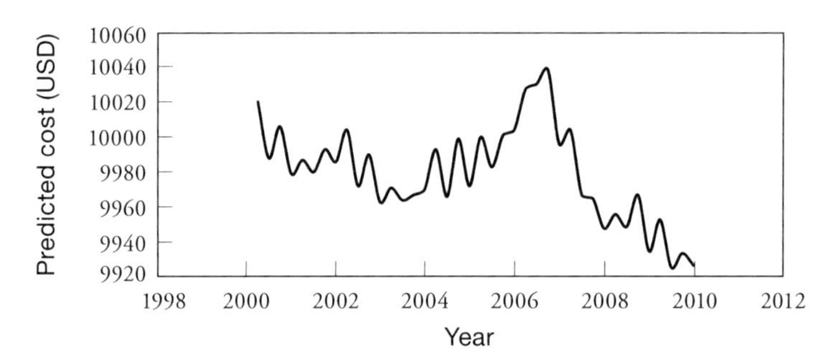

Figure 8.9 A broken scale.

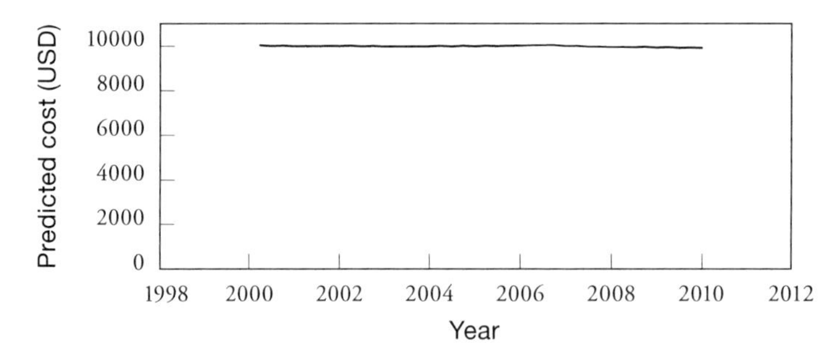

Figure 8.10 A true scale.

In systems analysis, it is wise to avoid broken scales because of their likelihood of deceiving the reader. When one must present a chart with a broken scale, it should be required that the scale be very clearly and prominently displayed to alert the reader. Similarly, the use of logarithmic scales can mislead, and, curiously, in some cases, the lack of logarithmic scales can be equally misleading. A "log" scale compresses differences, which for variables thought of as "linear" (e.g., money) can easily distort the impression a reader might gain. Some variables require a log scale to illustrate their "truth," for instance, the history of the Dow Jones Industrial Average. These concerns require systems analysts to be ever vigilant in their search for the truth in advertising.

8.8 Systems Analysis Tells a Story

> The shortest distance between a human being and truth is a story.
>
> —Anthony de Mello

The culmination of a systems analysis is the reporting out of the results: the story. And this is the most important part of all, for if the systems analysis is kept hidden away, it is of no use to anyone. We have discussed the essence of the analysis; now we will discuss the presentation. Of course, a story must have a beginning, a middle, and an end. One might say, "of course" (I did), but this three-part structure is not something to take for granted. Many are the stories reporting on systems analysis that have violated this command. As Steven Spielberg said, "People have forgotten how to tell a story. Stories don't have a middle or an end anymore. They usually have a beginning that never stops beginning."

The beginning tells what our problem is and what we are going to do. The middle tells what we did, and the end tells what it all means. In all three sections, we have several important concerns: responsibility, conciseness, and clarity.

8.8.1 Responsibility

We mentioned responsibility back in Section 8.5, but it is so important that it bears amplification here. The analyst has the responsibility to ensure that the story is true, and that the story is understood by the recipient.

We have spent some time on truth, but here it is well to emphasize that the need is for the story to *not be misunderstood.* The story, any graphic associated with the story, and the impressions created must not be used to tell a story at variance with the discoveries of the systems analysis. It is the responsibility of the systems analyst to achieve the true story.

In reporting the work, the systems analyst must tune the story to the audience. Too often a systems analysis is disregarded because the audience did not understand the conclusions or the evidence that led to the conclusions. We analysts all have heard the remark that our audience just "did not get it" as a criticism of the audience. It is, in fact, not really a criticism of the audience, but rather a criticism of the systems analyst. It is the responsibility of the systems analyst to tailor the story to the audience; the opposite is impossible.

Lastly, the systems analyst must be responsible for including exculpatory evidence, that is, evidence that is at variance with the story being told. The exceptions, the alternate choices, and the reasonable alternatives must be included to give a fair and balanced story to the audience. With a healthy regard for these responsibilities, the systems analyst can be satisfied that the story is true, not only as a fact, but in the eye of the recipient of the systems analysis.

8.8.2 Conciseness

Often, in telling our story, the hardest thing is telling the entire story in a short period of time. Telling a long rambling story is easy, and it is easy to put an audience to sleep. We are tempted to regale the audience about this interesting problem, the trials and tribulations we went through to answer it, the blind alleys we had to back out of, and the branches that led nowhere.

We do not need to tell everything we did to reach our conclusions, only enough so that a listener will understand what it was that led us to the conclusions. Sometimes, it helps to start planning our story with the conclusions, then work back to the introduction. As we do this, we must examine things along the way to assure ourselves that each step is absolutely necessary to provide evidence to support our conclusions. The important thing is the conclusions and the information supporting the conclusions, rather than the journey that led us there.

But, above all else, we must mold the story to fit into the time allotted, or it will fall on deaf ears. The strict rigor of sticking to our story, the decisive cuts in things that were so dear to us earlier, and the relentless pursuit of brevity are essential parts of telling our story and must not be neglected.

8.8.3 Clarity

The story we tell must be concise and exhibit simplicity. The reader then can see immediately the truths the systems analyst is expressing. In some cases, a single chart conveys the entire message. Two examples from vastly different venues can be used to illustrate this point: one so simple that it cannot be misunderstood, and one complex but with an elegance that, again, means it too cannot be misunderstood.

Figure 8.11 represents a comparison of a U.S. and an adversary's fourth- and fifth-generation fighters. Examples of these might be F-15s as fourth-generation and F-35s as fifth. We see U.S. fourth-generation fighters at rough parity with the adversary's fourth-generation fighters, yet at a severe disadvantage when they encounter the adversary's fifth-generation fighters. U.S. fifth-generation fighters suffer no comparable dilution of performance and provide a measure of superiority under all the conditions illustrated in the earlier analysis.

Figure 8.12 is the 1861 classic graphic by Charles J. Minard that illustrates an entire campaign, Napoleon's march on Moscow. The graph shows the geography spanned during the march; the changing number of soldiers at different points along the route, both to and from Moscow; and the ambient temperature (Russia's "secret weapon") along the route. Though the chart seems complex at first, on examination we see an elegant simplicity. The width of the tan line indicates Napoleon's force size in the march to Moscow; the black line width represents the retreating force size. The French forces were thoroughly depleted by the time they got to Moscow and were but a shadow of the "Grand Army" that had started. The declining temperatures further reduced the army such that virtually none of the army survived the disastrous Moscow campaign. I think this is one of the finest examples of a whole story in one picture.

These two illustrations were chosen to illustrate that there is not one template for the clarity in a systems analysis but a diversity dependent on the particular circumstances.

An additional challenge in presenting your story is that you may not get the amount of time you have been promised. So, you need even a shorter story. Have a story ready that goes in half the time and have a one-minute "elevator speech" ready.

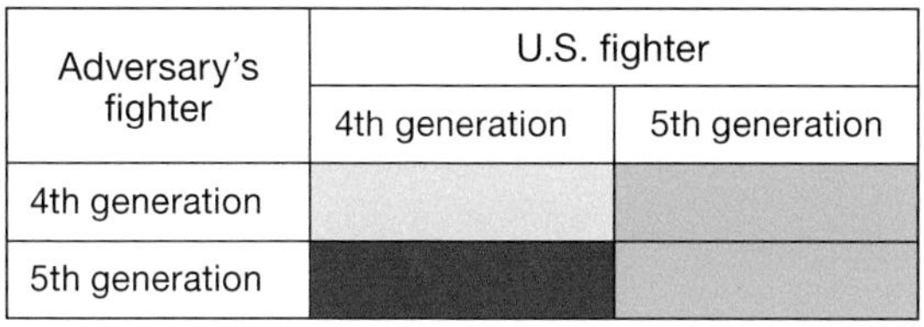

Adversary's fighter	U.S. fighter	
	4th generation	5th generation
4th generation		
5th generation		

Figure 8.11 Example of a concise story.

8.9 Conclusions

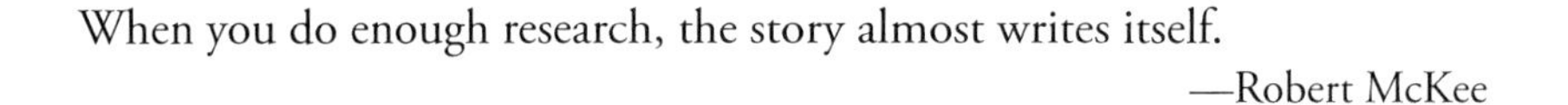

> When you do enough research, the story almost writes itself.
>
> —Robert McKee

I must admit that I have not done "enough" research to do justice to this chapter. For that reason, the story did not "almost write itself," and I have to conclude my story to keep it concise. This is, perhaps, a cautionary tale with some simple examples of where we as systems analysts can easily go astray.

Our overarching tenant is to be truthful. Recognizing that gives us a responsibility to tell not only the truth but the whole truth. The responsibility falls in having the math correct, a necessary but hardly sufficient condition for truth. The responsibility also involves telling the real or fundamental truth and, perhaps of equal importance, telling the story in such a fashion that it will be correctly understood and believed.

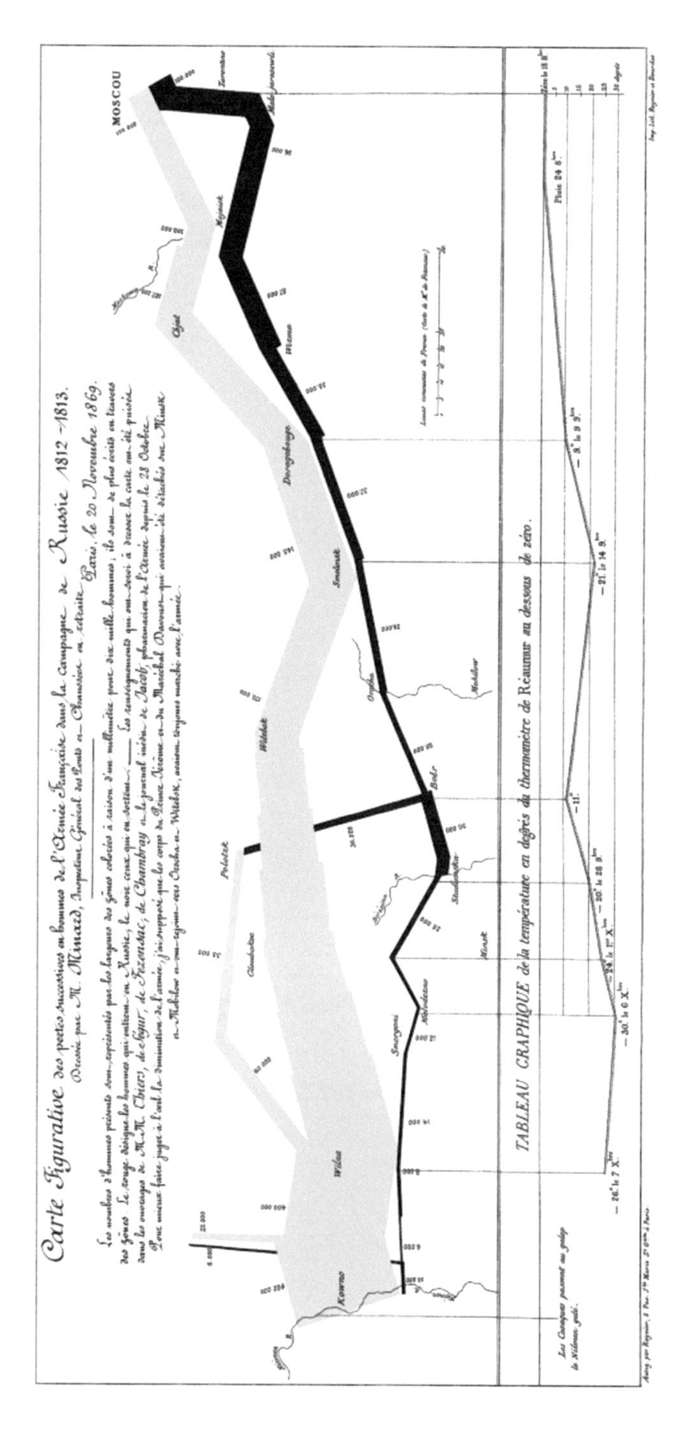

Figure 8.12 Napoleon's march. (Figure credit: Edward R. Tufte, *The Visual Display of Quantitative Information*, Cheshire, Conn., Graphics Press LLC, 1983, 2001.)

About the Author

Alan D. Bernard has been a Principal Laboratory Researcher at MIT Lincoln Laboratory since 2004. He is responsible for technology issues across the Laboratory, reporting to the Laboratory's director. Mr. Bernard's responsibilities include space systems, aeronautics, communications, and electronics.

From 2000 to 2004, he was the associate head of a division conducting research in technology related to the survivability and lethality of modern military aircraft. Earlier, as the leader of the Systems Analysis Group at the Laboratory, he directed 60 researchers in modeling, synthesizing, and evaluating various concepts for the insertion of advanced technology into modern weapon systems. He led the Air Force Red Team activity evaluating stealth, countermeasures, and capabilities.

Mr. Bernard joined the Laboratory in 1972. His activities in the Reentry Physics Group focused on modeling the interaction of reentering objects with the atmosphere and validating these models with flight-test data from the Kwajalein Missile Range. He developed and evaluated concepts for both ballistic missile offense and defense.

From 1963 to 1972, he was a staff scientist at the Avco Corporation. There he participated in the Apollo program, evaluating the thermal and kinematic performance of the command module's heat shield. He became flight-test project engineer in charge of electronic countermeasures and other aircraft penetration aids.

Mr. Bernard attended the University of Connecticut (BS, MS) and Boston University, majoring in physics with a secondary concentration in mathematics. He was elected to Sigma Pi Sigma, the national physics honor society.

III

Defense Systems Analysis in Specific Areas

9

Air Defense Systems Analysis

David J. Ebel

9.1 Introduction

In possession of an undergraduate degree in physics and a doctorate in mathematics, I probably was well prepared to be a systems analyst. Of course, I did not know what *systems analysis* meant when I joined Lincoln Laboratory in 1985. In fact, I did not think of the *process* much in my early work; I just worked on problem after problem, each problem seemingly new and different from its predecessor. I had become involved in the very dynamic Air Force Red Team (see Chapter 6 by Aryeh Feder), and I was very busy.

Years later, I found myself guiding newcomers in the process and began to think more about the protocols, procedures, and pitfalls of the systems analysis process. Along the way, I found that I thoroughly enjoyed the challenge of a new problem every few months, the opportunity to help clarify important problems in air defense, and the chance to interact with very interested, senior decision makers.

I hope my remarks here will be helpful to other practitioners of the art, particularly to newcomers to the field.

9.2 Systems Analysis and Air Defense

The objective of air defense systems analysis, as I practice it, is to gain an understanding of threats to U.S. air vehicles and to help design and acquire systems that enhance air vehicle survivability and effectiveness. Conducting systems analysis involves developing mathematical models that represent real-world warfare, performing computer simulations required by the models, and identifying the real-world implications of the results. By necessity, mathematical models and computer simulations greatly simplify the real world. Often the jump from the real world to the simulation domain is large. To bridge the gap, field and laboratory tests are used to validate models under simplified and controlled conditions. Model deficiencies can then be identified and corrected. Figure 9.1 illustrates the domains and their connections.

Some of my coauthors have argued that we may be able to go a long way in the analysis domain with just pencil and paper. My experience in air defense is probably similar to author Steve Weiner's experience in missile defense: "back-of-the envelope" calculations are an important first step, but the process is well defined enough to allow computer modeling to cover real-world scenarios more fully and more accurately.

Formulating real-world problems in a form suitable for computer simulation involves identifying key issues, deciding what to model and at what fidelity, creating figures of merit, and developing computer codes to do the modeling. Interpreting modeled results requires understanding the accuracy of the model, understanding its sensitivity to assumptions and unknowns, analyzing the impact of random or unmodeled effects, and ultimately using engineering judgment to determine whether an impact is negligible, modest, or significant.

A further challenge is illustrated by the air defense action and reaction cycle shown in Figure 9.2. Capabilities are not fixed in time; defense systems analysis needs to be concerned with near-, mid-, and far-term capabilities as well as current capability. A good systems analysis must consider the sensitivity of results to a range of potential

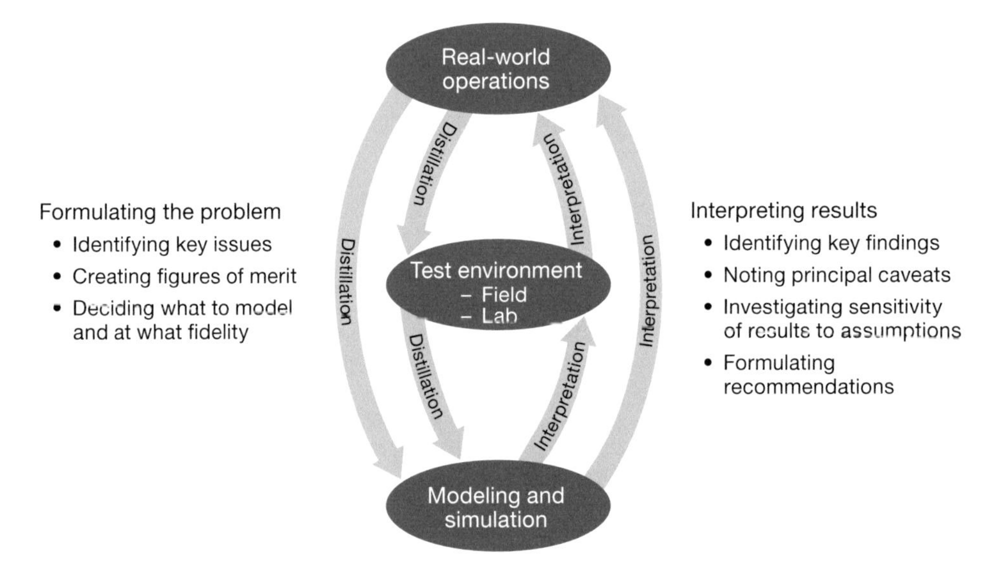

Figure 9.1 Systems analysis domains.

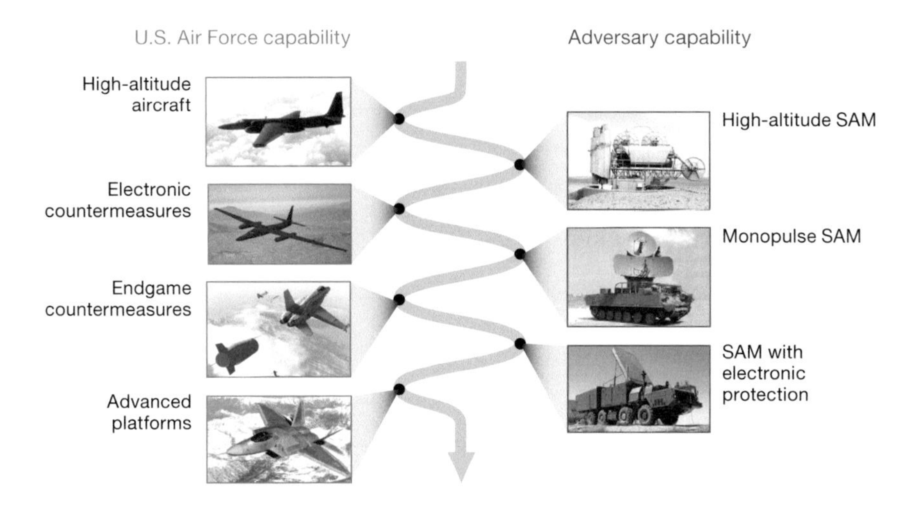

Figure 9.2 Air defense action-reaction cycle.

future threat reactions. The results can then be used to identify technologies and system architectures that are robust to evolving threat capabilities or that require threat development of expensive counter technologies.

9.3 Testing and Model Validation

The foundation of validated systems analysis is instrumented flight testing. Flight tests can be used to "check the box" to ensure that the system models provide a reasonably accurate answer. Frequently, however, surprises occur, and the test result diverges from what was expected. In these cases, the analyst can trace through the flight-test instrumentation to "debug" his system model and identify the phenomenology or physics that was missed. In other cases, a parameter is impossible to predict a priori, and test data provide the necessary empirical estimate. Examples of the latter are radar ground clutter and infrared background clutter. The analyst may find himself grappling with the infinite variety of the earth's surface or its environment in an attempt to mathematically model these effects.

In the 1980s, an extensive (heroic) radar ground clutter measurement campaign undertaken by Lincoln Laboratory aimed at understanding clutter phenomenology and developing models for estimating it.[1] Similar efforts have been undertaken for the measurement of infrared clutter. In the case of radar ground clutter, once the

1 J. Barrie Billingsley, *Low-Angle Radar Land Clutter*. Norwich, N.Y.: William Andrew Publishing– SciTECH, 2002.

phenomenology is established, it is a straightforward matter to model terrain visibility and to determine the grazing angle to that terrain. It is also possible to qualitatively identify the nature of the vegetation or other ground cover from available databases. The model development task then becomes one of creating mathematical models for effects that are amenable to calculation and developing empirical models for those that are not. Finally, the residual radar clutter after pulse-Doppler or moving target indication (MTI) processing can be estimated based on modeled input clutter and measured or known radar parameters, such as frequency stability, timing jitter, and antenna motion.

9.4 The Air Defense Engagement Chain

The starting point for any air defense systems analysis is to identify and model the elements of the engagement chain (or "kill chain" as referred to by some of our authors). A generic engagement chain that is useful for modeling integrated air defense systems is shown in Figure 9.3. The engagement chain can involve any combination of surface and airborne assets engaging other surface or airborne assets. The basic functions are the surveillance element where the target is detected and tracked, and a command-and-control element that fuses surveillance information from potentially a number of diverse surveillance sources and then allocates a fire-control element to engage the target. The fire-control element must search the uncertainty volume provided by the

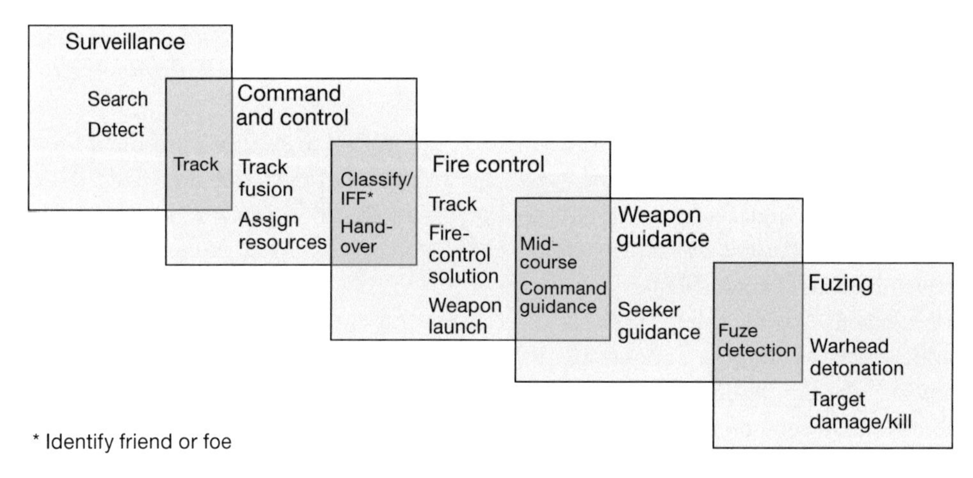

Figure 9.3 Air defense engagement chain.

surveillance and then guide a weapon to engage and fuze on the target, either autonomously or semiautonomously in conjunction with the fire-control element.

The envelope of overall air defense capability comprises a region or *footprint* in which the air defense system is effective. Each element of the engagement chain has its own footprint that can be convolved or otherwise mathematically combined with other elements to determine the footprint of the overall system. Thus, in order to determine the overall system footprint, the analyst must develop models for several sequential stages of detection, track, and handover between a potentially diverse set of system elements. The key parameters that determine the success or failure of handover between the successive elements are *detection range*, *track metric accuracy*, and *data latency* for each stage. Simply put, determining whether or not the engagement chain works involves determining whether a particular sensor is faced with a task analogous to finding a needle in a haystack or analogous to finding a needle from among a only few strands of hay. In the next three sections, we consider the tools that are most frequently used to make these determinations.

9.4.1 The Radar Range Equation

The basic tool for determining the detection range for each of the radio-frequency (RF) elements of the engagement chain is the radar range equation. It applies to surveillance radars, fire-control radars, missile seekers, and, with straightforward modifications, to other active and passive detection systems. The many variants of the radar equation all simply express the conservation of energy. Because this equation expresses the conservation of energy, test results must obey it! If they appear not to, then the search for an explanation involves carefully isolating each term in the equation and assigning error bounds to it. If that approach doesn't work, then a retest with better instrumentation and more carefully controlled test variables is in order.

I have spent many hours and days tracking down mistakes in using the well-known but much-abused radar range equation, so some advice is in order. Figure 9.4 shows the form of the radar range equation I favor (Steve Weiner in Chapter 10 derives the equation from basics and arrives at a variant form that is slightly different from mine). I like my form because its explicit elimination of bandwidth from the equation makes it is clear that bandwidth has no impact on radar detection (in the noise-limited case).

In using the radar range equation, it is important to note at what point in the signal processing chain the signal and noise terms are referenced. In the form presented here, the output signal-to-noise ratio is measured at the output of a matched filter and represents the peak signal-to-mean-noise ratio; the receiver noise power density is referenced at a different point: the antenna terminals. The utility of this approach

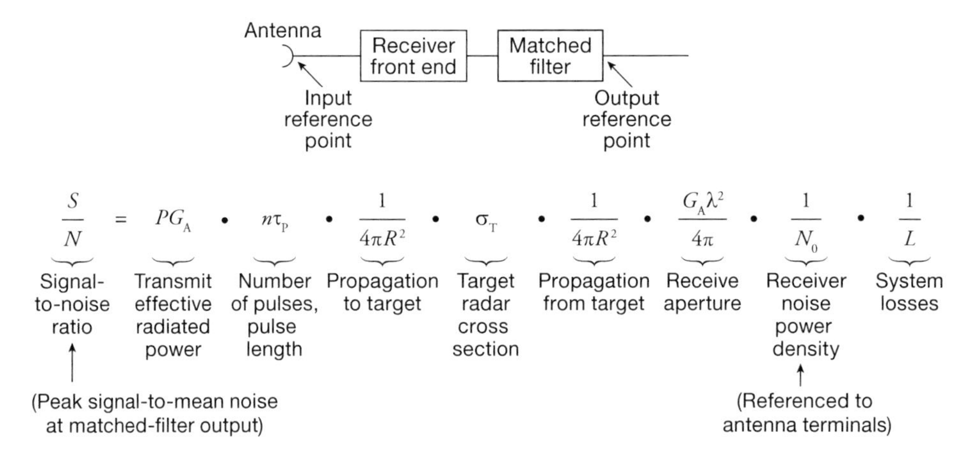

Figure 9.4 The radar range equation.

is evident if we now consider the impact of jamming on the radar detection range. In Figure 9.5, the radar equation is shown with jamming included. The J_0 term, or jamming noise power density, is easily calculated at the antenna terminals. At this point, J_0 is indistinguishable from receiver noise, and no additional signal processing losses should be applied to it. To apply additional signal processing losses leads to a double counting of losses on the jammer. The signal-to-interference or signal-to-($J + N$) at the output of the matched filter is correctly given by the radar range equation in Figure 9.5. (Test data must also agree with this form of conservation of energy.)

9.4.2 Metric Accuracy—Error Estimation

Formulae for estimating metric errors in a radar system are shown in Figure 9.6. The two most frequently estimated parameters are angle and range. Presented here is a simple heuristic derivation of the pulse time-of-arrival formula; the angle-of-arrival derivation is analogous. The slope of the arriving pulse envelope is seen to be approximately its amplitude divided by its half width. The effect of additive noise is to displace the pulse laterally by an amount equal to the amplitude of the additive noise divided by the slope of the uncorrupted pulse (for large signal-to-noise ratio). This formulation provides the error estimate shown in Figure 9.6 for the time-of-arrival estimator of the leading edge of the pulse. Because the back edge of the pulse provides an independent sample of noise, the overall formula has an additional dividing factor of root 2 to yield the final result.

$$\underbrace{\frac{S}{J+N}}_{\substack{\text{Signal-}\\\text{to-noise}\\\text{ratio}}} = \underbrace{PG_A}_{\substack{\text{Transmit}\\\text{effective}\\\text{radiated}\\\text{power}}} \bullet \underbrace{n\tau_P}_{\substack{\text{Number}\\\text{of pulses,}\\\text{pulse}\\\text{length}}} \bullet \underbrace{\frac{1}{4\pi R^2}}_{\substack{\text{Propagation}\\\text{to target}}} \bullet \underbrace{\sigma_T}_{\substack{\text{Target}\\\text{radar}\\\text{cross}\\\text{section}}} \bullet \underbrace{\frac{1}{4\pi R^2}}_{\substack{\text{Propagation}\\\text{from target}}} \bullet \underbrace{\frac{G_A\lambda^2}{4\pi}}_{\substack{\text{Receive}\\\text{aperture}}} \bullet \underbrace{\frac{1}{J_0+N_0}}_{\substack{\text{Receiver}\\\text{noise}\\\text{power}\\\text{density}}} \bullet \underbrace{\frac{1}{L}}_{\substack{\text{System}\\\text{losses}}}$$

(Peak signal-to-mean noise at matched-filter output)

(Referenced to antenna terminals)

Figure 9.5 The radar range equation with jamming.

9.4.3 Data Latency

Estimation of the effect of latency is the most straightforward effect for the systems analyst to model. Any data latency can be converted to an error in distance simply by using distance = velocity × time. For moving ground targets or airborne targets, time delays can be thereby converted to errors in distance and are then easily combined with the distance errors from other measurements. For fixed ground targets, any measurement, however old, is valid.

Using the three calculations for detection range, metric error, and data latency, or having models for them, the defense systems analyst can determine the probability of handover success at each successive stage of the kill chain and rough out an overall system performance model.

9.5 Model Hierarchy

Which calculations to emphasize and at what fidelity are two of the most difficult issues for newcomers to address. The answer to both is, it depends on the question being asked. If the problem to be modeled is the intercept of a hypersonic missile, for example, relatively simple models of the detection process may suffice. The analyst's efforts may be spent on developing detailed models of the interceptor missile guidance and of the various system latencies to ensure that the effects that matter most are modeled in highest fidelity. On the other hand, if the effect of a jammer is being considered, careful attention to modeling the radar electronic protection is warranted, and modeling the missile intercept with a simple flyout-time model may suffice. It is critically important to ensure that valuable computer execution time, as well as the analyst's modeling time, is directed at the problems that matter most.

Angle-of-arrival error

A B

Squinted beams

Angle-of-arrival estimator

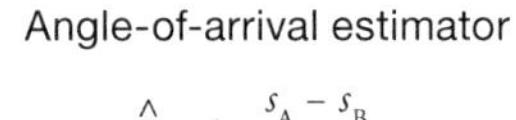

$$\hat{\theta} = k\,\frac{s_A - s_B}{s_A + s_B}$$

Angle error

$$\sigma\theta = k\,\frac{\theta_{beam}}{\sqrt{2\,S/N}}$$

where k is a constant of order unity, θ_{beam} is beamwidth, s_A and s_B are signal amplitudes from beams A and B, respectively

Pulse time-of-arrival error

Amplitude

Time

Early gate Late gate

Time-of-arrival estimator

$$\hat{\tau} = k\,\frac{s_E - s_L}{s_E + s_L}$$

Time-of-arrival error

$$\sigma_\tau = k\,\frac{\tau_P}{\sqrt{2\,S/N}}$$

where k is a constant of order unity, τ_P is pulsewidth, s_E and s_L are signal amplitudes from early and late gates, respectively

Derivation of pulse time-of-arrival error

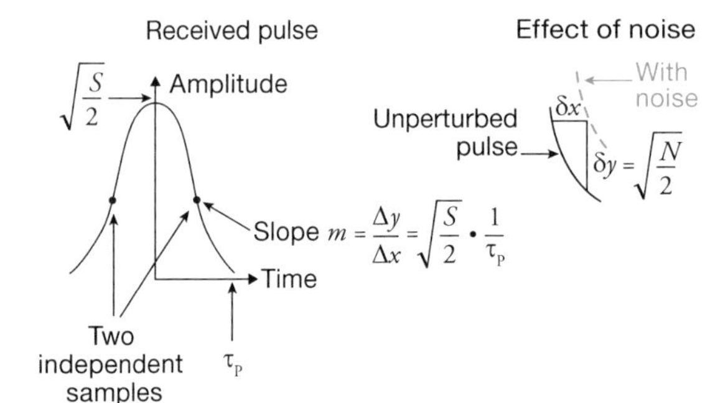

$$\delta x = \frac{\delta y}{m} = \frac{\tau_P}{\sqrt{S/N}} \quad \text{One sample}$$

$$= \frac{\tau_P}{\sqrt{2\,S/N}} \quad \text{Two samples}$$

Where δx is the time-of-arrival perturbation corresponding to a noise amplitude perturbation δy

Figure 9.6 Estimating metric errors.

Figure 9.7 illustrates a hierarchy of models and testing that are at the disposal of the systems analyst. The beginning point is the back-of-the-envelope calculation of the various subsystem elements in the air defense kill chain. Those calculations that fall near the borderline of subsystem success or failure can then be modeled at the next highest level. At some point, if there are uncertainties in the model, testing can be conducted to validate the model or refine an estimate of a parameter. The process can then continue up the modeling and testing hierarchies until it is clear whether

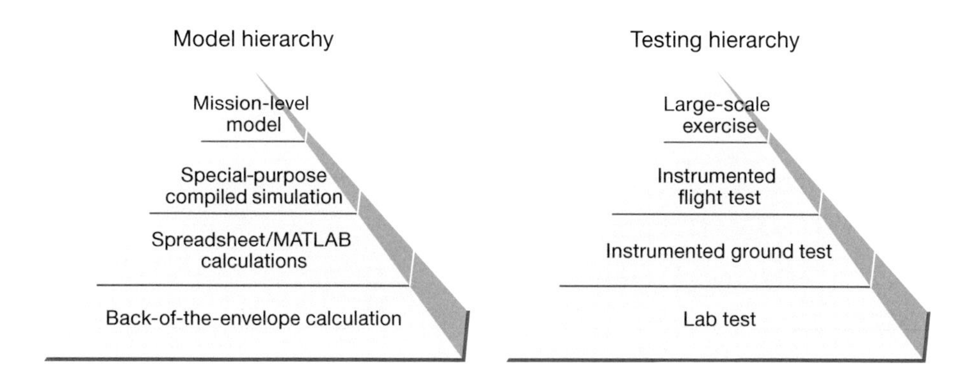

Figure 9.7 Systems analysis hierarchy.

the subsystem works or fails. As one migrates upward in the testing hierarchy, one encounters a steep increase in the cost of the test at each new level. Therefore, much thought needs to be given to getting the most out of the lower-level experimentation.

Once the analyst has addressed all of the major uncertainties, he or she can use a mixed approach, combining the lessons learned from each of the modeling and testing phases to provide an end-to-end assessment of system performance for a variety of scenarios and conditions.

Figure 9.8 illustrates how a hierarchy of models can be combined to make an overall assessment of the impact of a particular radar anti-jam technique called a sidelobe canceller. The sidelobe canceller concept relies on the fact that in the presence of noise jamming, most of the time-averaged power at the radar antenna terminals originates from the jammer itself rather than from the relatively few number of detected targets. Thus, if the output of a high-gain main antenna is combined with an adaptively weighted output from a low-gain auxiliary antenna so as to minimize the total power out of the system, the effect is to steer an adaptive null at the jammer. At the lowest level, this problem can be formulated mathematically as that of finding the minimum of a positive definite quadratic form by inverting the main and auxiliary beam covariance matrix. At the next level, a component model can be developed based on laboratory or field measurements of the main and auxiliary antenna patterns and of other radar parameters limiting cancellation. Finally, at the scenario level, component models for the radar subsystems, the jammer waveforms, and the target aircraft signature are used to determine the overall system footprint of the surveillance radar.

The sidelobe canceller is only one of many potential jamming countermeasures. In general, a combination of several approaches may be necessary. Figure 9.9 illustrates the "layered defense" approach to defeating jamming. The first objective would be to

Mathematical model

Average radar power output

$$P_{out} = \left\langle \left| s_{main} - w \cdot s_{aux} \right|^2 \right\rangle$$

where $\langle\,\rangle$ indicates average over time, and w is an adaptive weight

Finding the optimum weight

$$\frac{\partial P_{out}}{\partial \bar{w}} = -\left\langle s_{main} \cdot \bar{s}_{aux} \right\rangle + w \cdot \left\langle \left| s_{aux} \right|^2 \right\rangle = 0$$

$$w_{opt} = \frac{\left\langle s_{main} \cdot \bar{s}_{aux} \right\rangle}{\left\langle \left| s_{aux} \right|^2 \right\rangle}$$

Jammer cancellation ratio with optimum weight

$$\mathrm{CR} = 1 - \frac{\left| \left\langle s_{main} \cdot \bar{s}_{aux} \right\rangle \right|^2}{\left\langle \left| s_{main} \right|^2 \right\rangle \left\langle \left| s_{aux} \right|^2 \right\rangle}$$

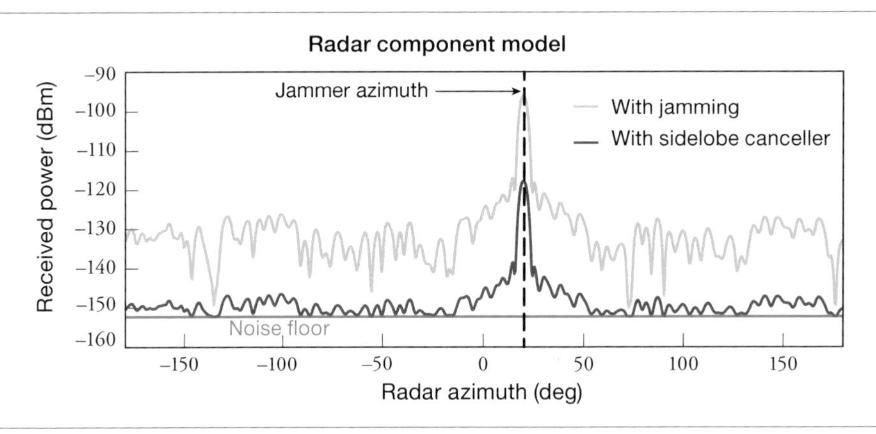

Scenario model

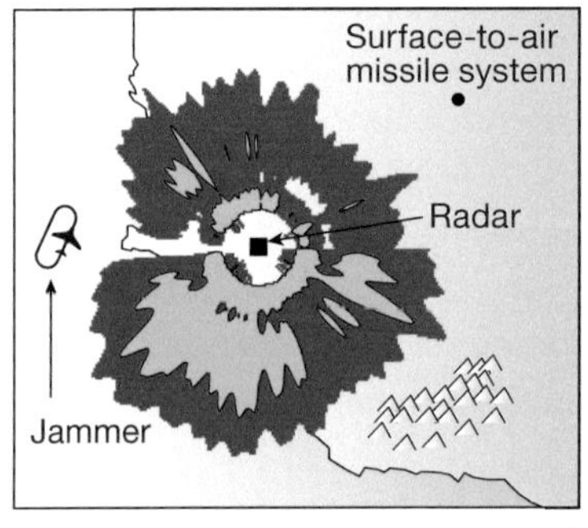

With jamming

With jamming and sidelobe canceller

Radar coverage footprint for notional 1 m^2 target

Figure 9.8 Three-level model hierarchy for a sidelobe canceller.

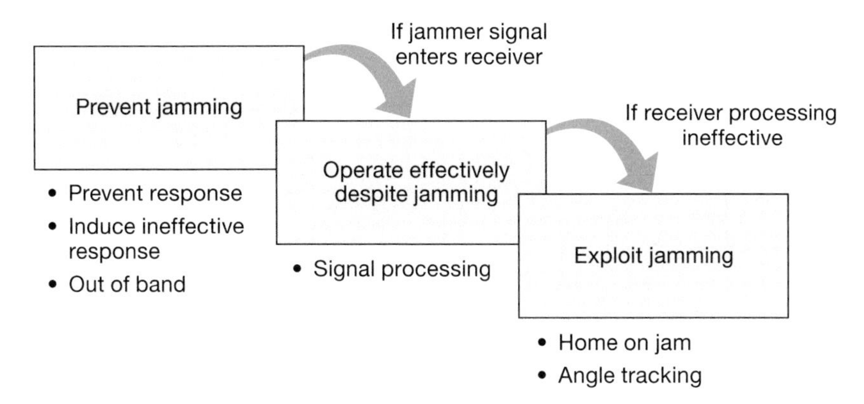

Figure 9.9 Jamming countermeasure options.

prevent the jamming signal from entering the radar receiver altogether. This challenge can be addressed in a number of ways: preventing the jamming response by defeating the jammer's ability to recognize the presence of the radar, preventing the jammer's signal identification function, inducing a delayed or ineffective response, or ultimately operating the radar out of band to the jammer. If these steps fail, signal processing techniques, such as sidelobe cancelling or sidelobe blanking, may reduce the amount of jamming signal entering the radar receiver, or matched-filtering techniques may allow discrimination of the desired signal from the jamming signal. If receiver-based techniques fail, it may be possible to exploit the jamming signal itself. This approach could take the form of angle tracking or triangulation techniques to locate the jammer and, ultimately, to destroy the jammer with a home-on-jam weapon. This last step of exploiting the jammer may carry a significant burden of training and tactics development for success. In considering a potential jamming technology or application, analysts will find it worthwhile to consider these categories of counters.

9.6 Fog of War

One of the most difficult issues to handle in defense systems analysis is the so-called fog of war, that is, the effect of unknowns or uncertainties faced during actual military operations. If either side in a conflict could identify the precise manner in which the fog of war would impact the outcome, they would likely take preemptive remedial action. As shown in Figure 9.10, the number of potential contributors to the fog of war is large, and hence, it is nearly impossible to a priori identify any particular contributor as the dominant issue. Figure 9.10 also illustrates the complementary nature of the capabilities of man and machine, and the way both contribute to the fog of war. For

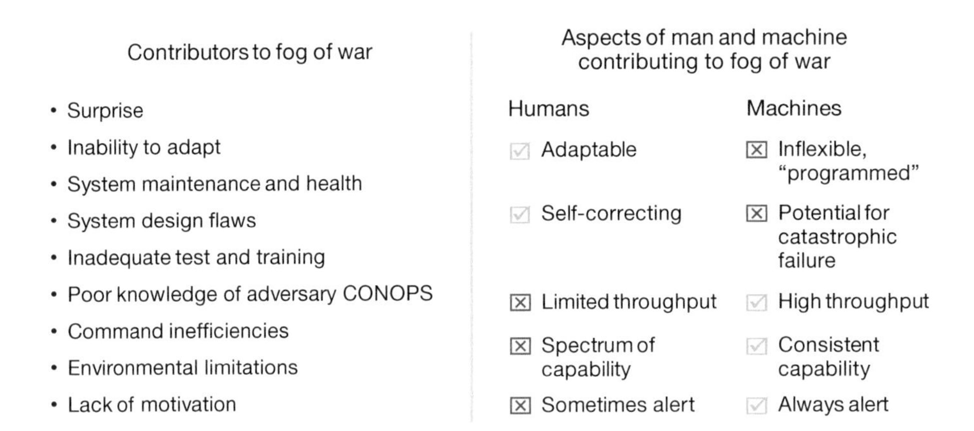

Figure 9.10 Contributors to the fog of war.

example, some forms of saturation attack or jamming might overwhelm the human's ability to keep up, whereas the computer could easily handle the throughput. On the other hand, if the attack possesses some unexpected characteristic, the human in the loop will adapt and look for workarounds, while the computer may fail catastrophically if the attack represents a situation not envisioned by the software architect. Today's trend toward increased sensing, computing, and automation clearly does not mean an end to the fog of war.

Part of the art of systems analysis is to factor out the fog of war from the major conclusions of the analysis. To determine the relative merits of, say, option A compared to option B, analysts can often assume the effects of fog of war to be the same in both cases. It then suffices to assume maximum technical capability on the part of the adversary. However, there are certainly cases in which the fog of war does not factor out; perhaps one option is more highly "tuned" than the other, requiring an exquisite knowledge of some part of the warfighting environment. In these cases, it is important to examine how gracefully the tuned approach degrades with decreasing knowledge of the environment. Handling the fog of war is one of many examples in which neither the warfighter nor the analyst can simply hand his job to the computer.

9.7 Case Study: Shepherds Flat Wind Turbine Analysis

In this section, we review as a case study the Shepherds Flat Wind Turbine Clutter Mitigation study undertaken by Lincoln Laboratory in 2010. The Shepherds Flat study illustrates an important air defense system design and modeling challenge: understanding the trade-off between maximizing radar target detection and minimizing radar false alarms. Example sources of radar false alarms are ground clutter, automobile traffic, thermal noise in a radar receiver, weather, and, as we shall see, wind turbines. The challenge in designing or operating a radar optimally is to define a target detection threshold that is sufficiently low to provide a good likelihood of detecting true targets (airplanes), while at the same time is not so low that the radar is flooded with undesired detections (false alarms). Because of the impossibility of predicting a priori all of the diverse sources of false alarms, most radars operate with an adaptive detection threshold referred to as constant false-alarm rate (CFAR) processing to maintain the total number of primitive target detections at some manageable number. Those detections are passed to subsequent data processing stages that form tracks and sort the desired true targets from the undesired false alarms.

The Shepherds Flat Wind Turbine Clutter Mitigation study began in April 2010, when Lincoln Laboratory was tasked by the Department of Defense to conduct a 60-day assessment of the impact of existing and proposed additional wind turbines on the Air Route Surveillance Radar (ARSR-3) at Fossil, Oregon (identification label QVN). Options for mitigating the impact of wind turbines on the QVN radar were

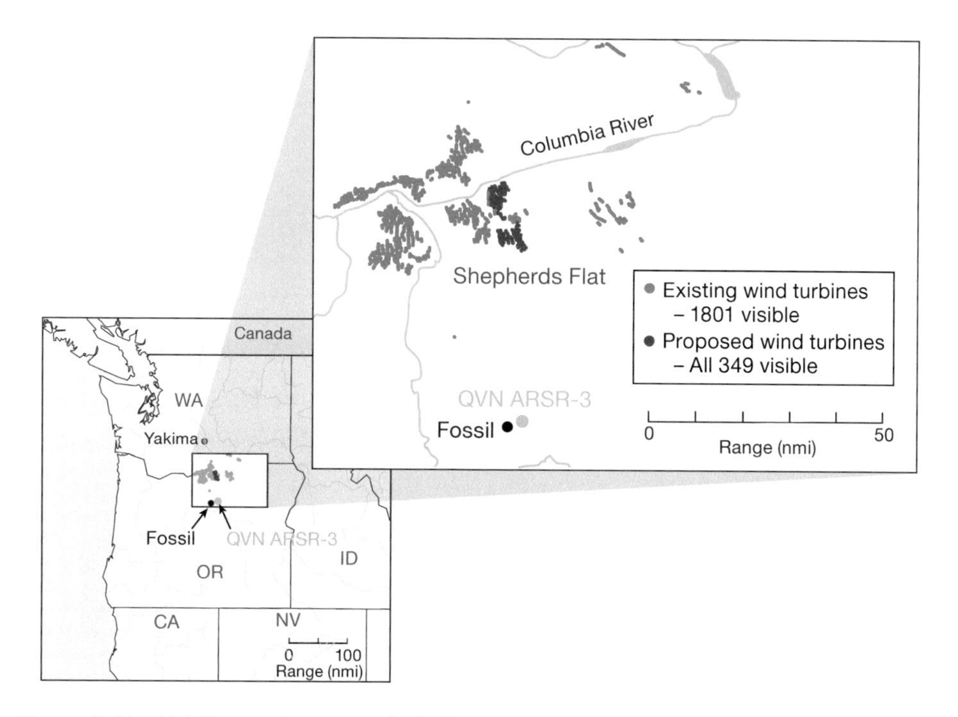

Figure 9.11 Existing and proposed wind turbines near QVN ARSR-3.

to be identified and evaluated. The map of Oregon in Figure 9.11 shows the general locations of existing and proposed wind turbines near the QVN radar.

Modern wind turbines can present a challenging radar clutter environment to a surveillance radar like the ARSR-3 shown in Figure 9.12. The principal source of

Figure 9.12 QVN ARSR-3 at Fossil, Oregon.

Figure 9.13 Wind turbines within line of sight of the QVN radar.

radar backscatter is the long blades (often 150 feet or more) of the turbines shown in Figure 9.13. A number of times in their rotation, the blades present very large radar returns (specular reflections), and these reflections can have substantial Doppler frequency components because the tips of these blades may move at speeds exceeding 100 miles per hour. These Doppler frequency returns are spread over a wide range in frequency, making them difficult to filter out in radar Doppler processors.

The proposal to add 349 more wind turbines at Shepherds Flat, near the Columbia River Valley, to an existing 1801 turbines caused serious concern to U.S. Air Force personnel operating the ARSR-3 radar on a mountaintop at Fossil, Oregon. They were concerned about the potential increase in the number of false alarms that would be caused by those turbines. Countering the Air Force's concerns were strong renewable energy and economic interests on the part of wind-turbine advocates.

Lincoln Laboratory was tasked to quickly assess the situation and to suggest remedies for any problems that appeared. We formed an assessment team consisting of three dedicated subgroups focused on modeling, field measurements, and data analysis. The modeling team constructed predictive performance models derived from ARSR-3 documentation provided by the Federal Aviation Administration (FAA) and radar measurements of wind turbines previously performed by the Air Force Research Laboratory. The field measurements team traveled to the QVN radar site with instrumentation for two test campaigns that included dedicated flight tests with a general-aviation aircraft. The field-test instrumentation package consisted of a multichannel, high-speed data recording system that allowed measurement of the native ARSR-3 signal processing, as well as an implementation of a modern digital "sidecar" receiver running in parallel with the native processor.

The data analysis team used the test data recorded from the QVN ARSR-3 to validate the performance models and evaluate candidate mitigation options. Before the field measurements began, it was not at all certain that they could be successfully conducted in the time allotted. However, the instrumentation package performed flawlessly, and upon completion of the second measurement campaign, it was clear that the field-test data collected greatly exceeded expectations.

Since the QVN ARSR-3 already had some 1800 wind turbines within its coverage area, the existing wind turbines provided a gold mine of useful data for determining the impact of wind turbines on the radar. The multichannel data provided the foundation for a characterization of turbine impacts on the ARSR-3 and, further, enabled a detailed quantitative analysis of potential mitigation options. In the end, very little was required of the modeling effort except to fill in a few gaps remaining after the data analysis.

The instrumentation data showed that QVN ARSR-3 radar performance is impacted by clutter resulting from many sources in addition to wind turbines— for example, road traffic, terrain, birds, and precipitation. These clutter false alarms are only partially mitigated by the ARSR-3 native clutter-filtering capability. The number of false targets generated by wind turbines accounted for approximately 10% to 20% of the total produced by the QVN radar, as shown in Figure 9.14. Road traffic, birds, and other clutter sources dominated. However, the existing 1800 wind turbines within line of sight of the QVN radar occupied only approximately 2.5% of the total terrain visible to the radar, so "pound for pound" turbines are clearly one of the most challenging clutter sources.

Analysis of the data collected showed that the number of false targets generated by the proposed wind farm, as well as those from the existing wind farms within line of sight of the QVN radar, could be reduced by proper adjustment of the radar settings. The study provided specific recommendations for this optimization. Analysis of archived QVN detection data, coupled with the pulse-level field-test data, showed that the overall false target count could be further reduced by modifications to the QVN radar. An auxiliary processor could be added to implement a modern, adaptive

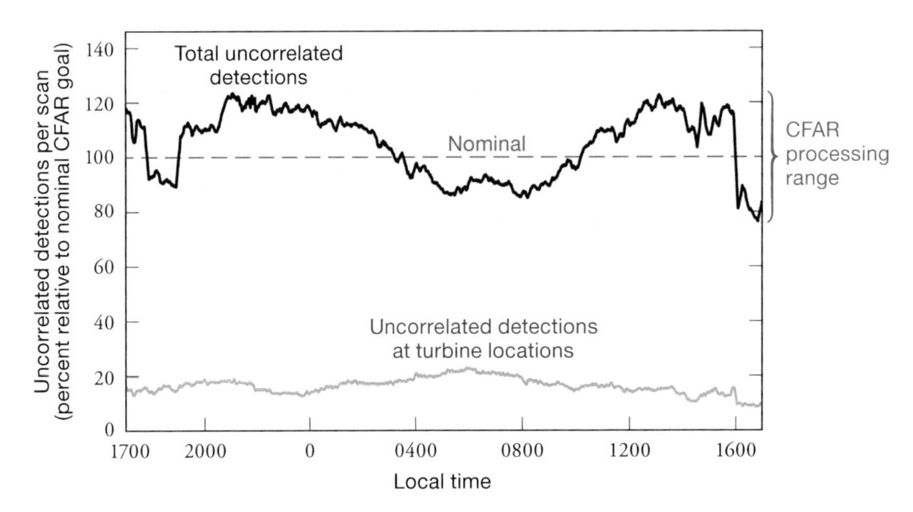

Figure 9.14 False alarms for QVN ARSR-3 radar.

clutter map that would edit out the false targets to prevent them from being passed downstream to display systems. Shortly after completion of the study, the FAA and U.S. Air Force implemented the recommended optimization of the QVN radar settings. The resulting performance improvement was determined to be sufficient that further modifications to the radar would not be necessary.

The major conclusions of the study were obtained from direct analysis of high-speed instrumentation data from the QVN ARSR-3 radar itself. The data collected allowed precise quantification of the trade-off between wind-turbine false-alarm mitigation and target detection; further, the data enabled identification of particular radar settings in need of adjustment. The Shepherds Flat wind-turbine study provides an excellent example of how instrumentation-quality test data can obviate the need for detailed electromagnetic and propagation modeling with its inherent uncertainties.

9.8 Parting Advice to the Aspiring Systems Analyst

In closing this chapter, I would like to summarize with some simple guidelines that I have learned from years as a practitioner of the art of systems analysis:

1. Know the question you are trying to answer and use it to guide your analysis. Don't waste brain cells, cups of coffee, or computer CPU cycles on the wrong problem.

2. Perform an end-to-end, first-order analysis; find the key drivers. If you are unsure how something works, make an estimate; first-principle physics is a good guide. Engineering systems usually don't depart radically from what physics or computation suggests.

3. Investigate key drivers with higher-fidelity models. Use the first-order analysis to identify system elements that merit a higher-fidelity modeling effort. Find the right balance between breadth and depth. Sanity check the high-fidelity results by comparison with first-order expectations. In the end, high-fidelity computer runs are an aid to understanding; they are *not* "the answer." In short, don't hand off your real job as analyst to the computer.

4. Address the evolving future; consider system robustness to future mitigations or counters. Avoid letting the side with the last move win.

5. Make your report or briefing concise, clear, and transparent. Tell an interesting story and deliver it on time.

About the Author

David J. Ebel is associate leader of the Systems and Analysis Group at MIT Lincoln Laboratory. Since joining the Laboratory in 1985, he has been an active member of the U.S. Air Force Red Team, a 30+-year collaboration between the Air Force and Lincoln Laboratory. The Red Team provides independent analysis of air vehicle survivability and air defense technologies to Air Force and Department of Defense decision makers. Dr. Ebel holds a BS degree in physics from Iowa State University and a PhD degree in mathematics from the University of Wisconsin.

10

Ballistic Missile Defense Systems Analysis

Stephen D. Weiner

Surgeon General's Warning:

This chapter contains a detailed treatment of ballistic missile defense systems analysis.[1] Missile defense admits of a fairly detailed analysis process because the objectives and capabilities of the offense and the defense can be relatively well understood.

Hundreds, even thousands, of missile defense systems analyses have been done since their start with the Nike Zeus system in the late 1950s. Thus, this treatment of defense analysis is more comprehensive than other chapters of this book, which treat problems that are generally less well defined. The reader should feel free to skim through the chapter, slowing down for specific topics of interest. If your ambition is to be a missile defense analyst, you had best read and understand it all.

10.1 Becoming a Systems Analyst

I joined Lincoln Laboratory in 1965 after finishing my PhD program at MIT. I was fond of referring to myself as a "low-energy physicist." Once I joined a group that was working on advanced radar techniques for ballistic missile defense, I became involved in some interesting investigations of wideband radar electromagnetic backscattering from missile offense threat objects, and I got a chance to use some of my physics background.

1 All Ballistic Missile Defense System (BMDS) and threat information contained in this chapter is notional and not representative of actual Missile Defense Agency (MDA) BMDS weapon systems or sensors' capability and performance. The opinions expressed are the author's own and do not represent the views and opinions of the Department of Defense (DoD), and should not be interpreted as such.

The group I joined had two somewhat loud and flamboyant staff members who were doing broad analyses of missile defense systems and architectures, Joel Resnick and Bill Delaney. They seemed to be having a fun time with their work and were enjoying a lot of visibility from higher-ups at the Laboratory and in Washington. Along the way, they were joined by a third character, John Fielding, who fit right in with their "fast-gun" style.

The analyses they were doing seemed pretty basic to me, and I was surprised that these analyses had not already been done by the Army folks who ran missile defense and their technical agent, the famous Bell Telephone Laboratories.

I was drawn into supporting their work without realizing that I was setting a 45-year course in my career doing missile defense systems analyses. My 45 years were fueled by a seemingly endless chain of changes in the nation's goals and desires for missile defense, and by a technology of offense and defense that was improving rapidly. At Lincoln Laboratory (as at Bell Laboratories), we had a great source of real data coming to our laboratories from essentially weekly full-scale missile tests launched from the United States and impacting in the nation's test site at Kwajalein in the Marshall Islands. These early years were an exciting mix of analyzing proposed systems, generating new architectures, and trying to understand all the ramifications of our testing in the Pacific. The debate about missile defense continued through many subsequent eras, and there has always been plenty to do. Along the way, I have made a good living from systems analysis.

10.2 Introduction

Ballistic missile defense (or BMD) has similarities and differences with the other subjects covered in this book. It is probably the topic best suited to mathematical game theory analysis. At a basic level, it consists of two moves: the offense selects an attack consisting of a number of missiles possibly carrying penetration aids sent to a variety of defended targets, and the defense responds to this attack with sensors to track and identify the threat objects and with interceptors to destroy the selected objects. BMD has all the essential features of a game theory problem. The offense and the defense have diametrically opposed objectives. Neither the offense nor the defense has full knowledge of the opponent's capabilities or plans. Furthermore, there are random elements in the battle that neither side can foresee. Examples of such randomness include missile reliability, countermeasure effectiveness, sensor measurement errors, and interceptor reliability. Both sides must account for this randomness as well as the opponent's strategy in making their plans.

In spite of all this uncertainty, BMD systems analysis is largely a straightforward process. For much of its history, BMD has been concerned with defense against missiles carrying thermonuclear warheads attacking major cities. As such, the definition of defense success and failure is pretty simple. If a missile gets through, the defense has lost. If all the missiles are stopped, the defense has won. In analyzing BMD performance, we generally consider two measures of effectiveness: coverage and leakage. These measures are the quantity and quality of the defense we provide. We try to maximize coverage and minimize leakage. Both measures can be improved by spending more money, and much of BMD analysis focuses on trying to meet coverage and leakage requirements at minimum cost.

Both coverage and leakage are amenable to fairly simple mathematical analysis. We often say that the two basic equations of BMD require only algebra or arithmetic. Coverage is usually analyzed using a timeline that looks at the sequence of functions the defense must perform and locates where and when these functions can be carried out in space and time. One of the basic equations for these coverage calculations is

$$\text{Distance} = \text{Speed} \times \text{Time}. \tag{10.1}$$

If we look at an attacking trajectory, there is an earliest time at which the defense has detected and identified a target, and tracked it accurately enough to assign an interceptor to it. There is also a latest time at which the target can be intercepted without causing any damage to the defended area. The distance that an interceptor can fly in the time between the earliest launch and the latest intercept determines the defense coverage. This calculation depends on the average speed of the interceptor. If this coverage is not adequate to cover the defended area, we need to deploy more interceptor sites, build faster interceptors, arrange to launch them earlier, or use some combination of these approaches. Much of BMD systems analysis is an attempt to find the most cost-effective approach to obtaining the needed coverage.

The analysis of leakage is governed by an equally simple equation:

$$\begin{aligned}&\text{Probability of success} = \\ &(\text{Detect probability}) \times (\text{Identify probability}) \times (\text{Kill probability}).\end{aligned} \tag{10.2}$$

The leakage probability is 1 minus the probability of success. If any of the key defense functions fail, the overall engagement will fail. Again, the defense can reduce leakage by providing redundancy in one or more of its functions or by improving the performance of a function by buying better sensors or interceptors. We will examine what goes into these calculations in the rest of this chapter.

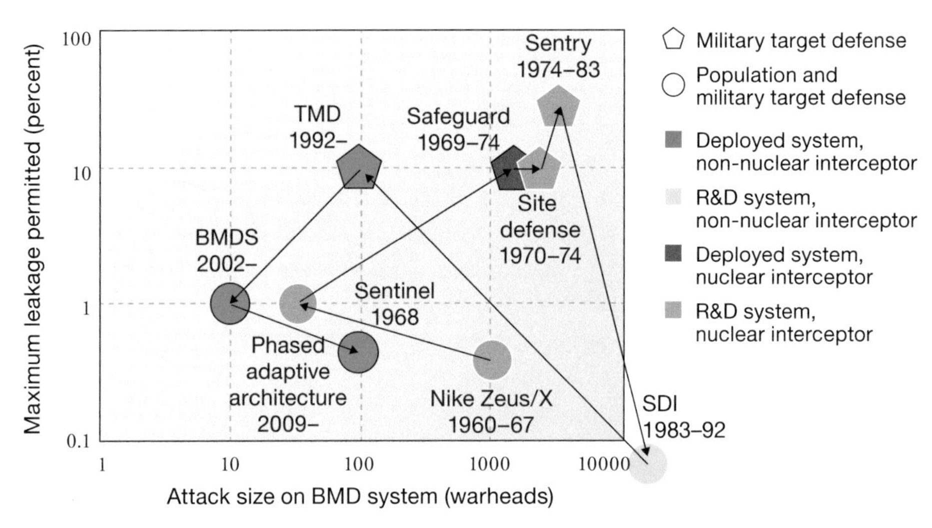

Figure 10.1 Evolution of BMD missions and systems.

Before we get into the details of various BMD systems analyses, it is worthwhile to review the history of the U.S. BMD program over the past 50 or so years. Figure 10.1 plots various BMD systems or concepts in a two-dimensional space measured by the attack size to be handled and the maximum leakage acceptable. The attack size is an offense parameter while the leakage is a defense parameter.

In this figure, systems to defend military targets are marked by pentagons while those defending the population as well as military targets are marked by circles. Deployed systems are marked by saturated colors while research and development (R&D) systems are marked by pastel colors. Finally, systems using interceptors with nuclear warheads are marked in red while systems using non-nuclear interceptors are marked in blue. The background space is colored to indicate the degree of difficulty of the BMD job, ranging from red for systems having very low leakage against very large attacks to green for systems that can tolerate high leakage against small attacks. The time evolution of BMD systems represents the changing balance between offense and defense, and also the balance between optimism and realism.

The initial Nike Zeus and Nike X systems were designed to protect large U.S. cities against a large Soviet attack. As the Soviet threat developed, it was realized that it was not possible to achieve low leakage against a massive attack, and furthermore, these systems could not cover the whole country, leaving many small and medium-size cities undefended. At this point, our defense posture shifted to one of deterrence; we could not stop a massive Soviet attack, but we could retain the capability to deliver a devastating retaliatory response even if we were attacked first.

In the late 1960s, China developed nuclear weapons and ballistic missiles, and the Sentinel system was proposed to defend against this much more limited attack. This problem was more manageable, and a long-range exoatmospheric interceptor, the Spartan, was developed such that about a dozen interceptor sites could cover the whole country, including Hawaii and the populated portions of Alaska. Before Sentinel could be deployed, further analysis of its limited ability to handle exoatmospheric penetration aids, delays in evolution of the Chinese threat, and further growth in the Soviet threat (combined with a change in administrations) led to a reconsideration of this deployment. The focus of BMD shifted to a potential Soviet threat to our deterrent forces and, in particular, to an attack by accurate intercontinental ballistic missiles (ICBM) on our Minuteman ICBMs. It was seen that a BMD system defending our Minutemen could negate this threat to our deterrence. The Sentinel components (radars and interceptors) would be deployed to defend Minutemen in a system renamed Safeguard. As seen in the figure, although the attack size for Safeguard was beyond what the Nike systems could handle, the acceptable leakage would be more than an order of magnitude larger. This was because if only 20–30% of the Minutemen survived, the remaining ICBMs would still constitute an effective retaliatory force. Furthermore, since the Minutemen were deployed in limited geographical areas, the coverage required of Safeguard would be much smaller than that needed for Sentinel. Although Safeguard was not designed for defending ICBMs, it was deployed, and R&D continued on dedicated follow-on systems called Site Defense and Sentry. Sentry was perhaps unique in BMD history in that it could have a leakage as high as 30–50% and still be successful.

All this changed with President Reagan's introduction of the Strategic Defense Initiative (SDI) in 1983. As seen in Figure 10.1, SDI had the objective of defending the entire United States against a massive Soviet attack on population, with the further constraint that the interceptors not use nuclear warheads. SDI was a research program looking at defenses operating in all phases of the threat trajectory: boost, midcourse, and reentry. It looked at defense sensors and weapons based in space, in the air, and on the surface. Although we will look at some of the analysis performed for SDI, there were never any plans to deploy the system. One lasting result of the SDI program was the shift from nuclear to non-nuclear interceptors. All subsequent BMD systems have used non-nuclear interceptors. This shift has required advanced technology for interceptor seekers and guidance, but the technology has been up to the challenge.

After the Gulf War in the early 1990s, the emphasis in BMD shifted to theater missile defense (TMD). The Army and Navy developed enhancements to the Patriot and Aegis air defense systems to handle short- and medium-range ballistic missiles. The Patriot was used against Iraqi short-range missiles with mixed success. Since these missiles did not use weapons of mass destruction (WMD), it was difficult to assess the BMD success. The damage done by the impact of a heavy, high-speed missile is similar whether or not a high-explosive warhead is detonated. If these missiles had carried WMD warheads, the BMD success rate would have been more obvious.

In the late 1990s and early 2000s, North Korea developed and tested longer-range missiles and were predicted to shortly have an ICBM that could reach Alaska, Hawaii, and eventually the continental United States. The requirements for a defense against this threat again pushed us into the low-leakage, small-attack-size portion of Figure 10.1. This system called the BMD System (BMDS) was deployed in 2004 and is still operational. Currently, there is work to extend this system to provide defense of our friends and allies against an attack with medium-range missiles. Here we will have to handle larger attacks while maintaining low leakage. Against shorter-range missiles, it is possible to use smaller (cheaper) sensors and interceptors. The current concept is called the Phased Adaptive Architecture, and its ultimate configuration is still evolving.

A common theme in the evolution of BMD systems and technology over the years is that we are always hopeful the BMD problem of the moment is solvable with current or near-term technology. Originally, we tried to deal with the most difficult case of a massive Soviet attack on population but realized it was too difficult (we revisited this problem in the SDI era with the same conclusion). However, over this window of time, we were presented with important problems that appeared easier to solve. These included defense of population against limited attacks, first by China and later by North Korea or Iran, and defense of military targets for which higher leakage would be acceptable.

In the rest of this chapter, I show some sample analyses that were done for many of the systems discussed above. For some of the analyses, the focus was on achieving large coverage at low cost. For most of the analyses, the focus was on reducing the leakage resulting from a variety of sources. For almost all the BMD systems considered, a major source of leakage was the ability of the defense sensors to identify the targets to be intercepted. This function is called *discrimination* and is one of the greatest uncertainties in a BMD system. I describe several approaches to characterizing discrimination performance.

The following sections have a matrix organization in which we look at the analysis of different aspects of BMD systems (coverage, leakage, and discrimination) across the major BMD missions (population defense, silo defense, and theater defense). The sections cover the defense functions or aspects, and specific examples cover the BMD missions. Finally, the chapter concludes with a section covering some general principles of systems analysis with specific application to BMD systems analysis.

The next few sections have some mathematical parts that are not essential to the narrative. The verbal description of what the math tells us and the sample results should give you an understanding of what we are trying to analyze. The equations sketch out the mathematical approach to the analysis problems and could give a head start to those readers interested in reproducing the results, although in many cases considerable additional math has been omitted.

10.3 Coverage Analysis

This section describes how to calculate the coverage of a BMD system. First shown is how we can trade off sensor range and interceptor speed to achieve a given defended region. Next, we look at the individual components—the interceptors and the sensors. In the interceptor section, we see how the interceptor coverage depends on its speed and the time it has to fly to the intercept point. In the sensor section, we look at how sensor coverage is limited by geometry (horizon issues), geography (where we can locate), and design issues. We derive equations relating how big or expensive a radar or infrared sensor is to how far that sensor can see. If you don't want to go through the details of the derivations, all you need to know is that to get a large coverage area, you have to spend a lot for big interceptors and big sensors.

What factors determine the coverage of a BMD system? Sometimes, we are trying to defend specific points such as missile silos, ships, or military bases. At other times, we are trying to defend an entire country. We have considered point defense systems that are sized to defend a single point. We have considered area defense systems that are sized to defend every point in a given region. These area defense systems can also be used to cover a discrete set of points. The area that can be defended by a BMD system is called its coverage or its footprint. The following simple analysis illustrates the major issues and indicates both the simplicity and the complexity of the analysis. Figure 10.2a shows a simple engagement with a defense site comprising a sensor and an interceptor defending a point uprange from the site. If this point is the furthest uprange one that can be defended, it is called the forward footprint.

This example is appropriate for short-range missiles for which we can assume approximately straight line trajectories and a flat Earth. Later, we see how things change for longer-range missiles.

The BMD system coverage depends on both the interceptor coverage and the sensor coverage. The next subsections look at these individually.

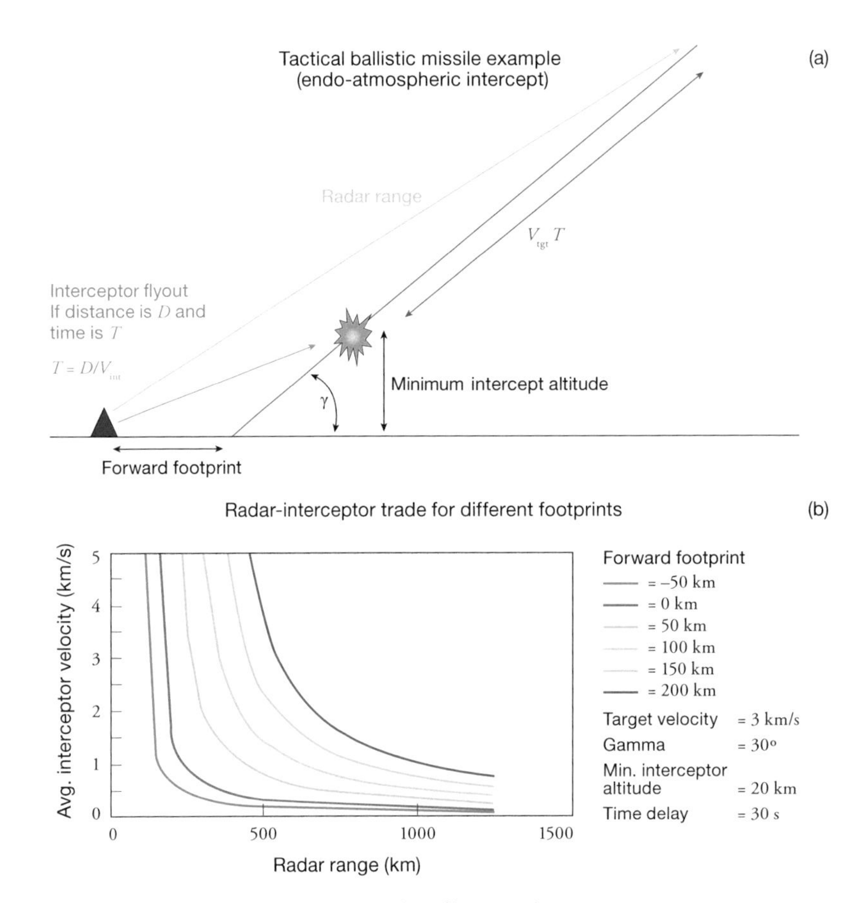

Figure 10.2 Radar-interceptor trade-off example.

10.3.1 Interceptor Coverage

We generate the trade-off curves in the lower part of Figure 10.2 by working backwards from intercept. We find the impact point corresponding to the desired forward footprint and look at a trajectory impacting here with a given velocity of the target (V_{tgt}) and reentry angle (γ). This trajectory is the red line in Figure 10.2a. We have a minimum intercept altitude determined either by the interceptor capability or by the need to avoid ground damage should the attacker detonate just above the intercept altitude. We then calculate the time it takes for the interceptor to fly to this intercept point. For this approximation, it is just the distance to the intercept point divided by the average interceptor speed. This time is represented by the blue line in

Figure 10.2a. We add some time for detection, discrimination, and decision making and see where the target was at this time prior to intercept. We then calculate the sensor range needed to see the target at this earlier time. This range is represented by the green line in Figure 10.2a. We repeat this calculation for a range of interceptor speeds. This process generates the trade-off curves shown in Figure 10.2b. We see that we can achieve a given forward footprint with a faster interceptor and a shorter range sensor or vice versa. To achieve a larger footprint, we need either a better sensor or interceptor (or both). Note also that we are still interested in negative forward footprints. This situation corresponds to a defense battery that cannot defend itself but can defend targets behind it, and is of interest for mobile defenses in which the attacker may not be able to attack the defense unit itself.

One important application for such trade-offs is to minimize defense cost. Longer-range sensors cost more than shorter-range sensors, and fast interceptors cost more than slow interceptors. If we have to defend a large area, we will need more defense units having small footprints than we will need for the case with larger footprints. These trade-offs are basically quantity-quality ones that will occur many times in BMD systems analysis. Yet another consideration that goes into these trades concerns the size of the attack. In general, a sensor can be reused for all missiles in the attack while an interceptor can only be used once for one attacker. If the interceptors were perfect, we would need one for each attacker, so for an attack of N missiles, we would need one sensor and N interceptors. Depending on the value of N, the minimum cost point on the trade-off curves in Figure 10.2b might tend toward the slower or faster interceptor regions.

The above type of analysis is appropriate for BMD systems defending point targets or small regions, as would be the case for silo defenses or military target defenses. Those types of BMD systems typically have the sensors and interceptors collocated, and cost is a very important consideration. For defending large areas, as is usually the case for population defense, the analysis becomes more complicated, and the interceptors may be very far from the sensors that provide their targeting information.

For long-range attacking missiles, this analysis must be modified to account for curvature of the Earth and also curvature of the target and interceptor trajectories. This analysis is more complicated but essentially the same as that shown in Figure 10.2. We briefly indicate the process and some of the new features of BMD that emerge.

For defense against long-range missiles, it is usually desirable to have the sensors and interceptors in different locations. The following analysis shows why this is the case. We can decouple the interceptor analysis from the sensor analysis. First we look at the interceptor coverage. Figure 10.3 indicates the time it takes the interceptor to fly to any point within its reach. This plot is usually referred to as the interceptor flyout plot and is rotationally symmetric about the vertical axis.

The upper plot of Figure 10.3 indicates the flyout time contours for a long-range interceptor. A maximum altitude and maximum ground range are within the interceptor coverage. In general, it takes longer to get to the edge of the coverage than to get to the top. In the lower plot of Figure 10.3, we superimpose the attacking trajectory in the same interceptor-centered coordinates. In this case, there is a range of possible intercept point locations. For each potential intercept point, there is an interceptor flyout time and a corresponding interceptor commit time when the interceptor needs to be launched in order to get to the intercept point when the target gets there. (This type of calculation is fairly simple for ballistic missiles because they do not generally maneuver outside the atmosphere.) The interceptor needs to be told where to go to hit

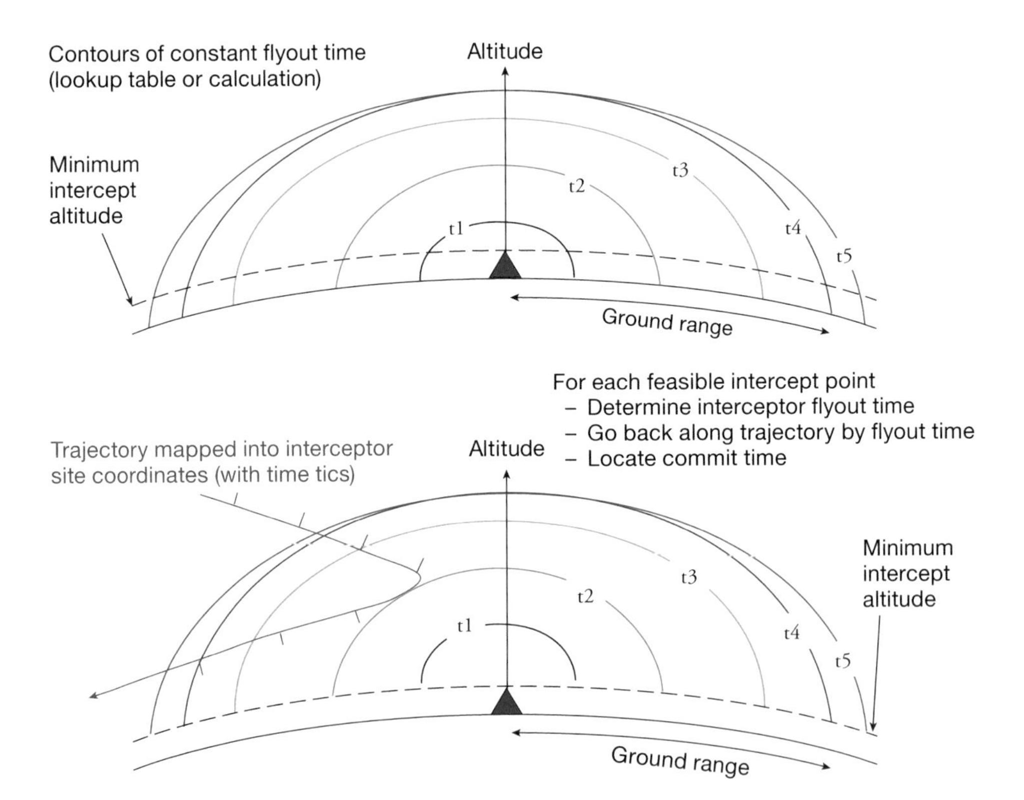

Figure 10.3 Interceptor flyout.

the target, and that is the job of the sensor network. These sensors need to provide a sufficiently accurate prediction of the target trajectory before we are willing to launch the interceptor.

In many cases, the target missile flight time is long enough that the defense may have multiple shot opportunities and the chance to replace interceptor failures. This process is called shoot-look-shoot (SLS) and offers the potential for significant savings in the number of interceptors needed to achieve a given level of performance. The analysis shown in Figure 10.3 can be extended to cover these SLS cases, as shown in Figure 10.4

Here we consider one particular ICBM trajectory from the Middle East to Boston. There is a long-range interceptor site in North Dakota. The table in Figure 10.4 indicates when we are able to launch an interceptor (commit window) and when an intercept can occur (intercept window). If we consider intercepts at different points along the trajectory, the black curve in the top plot shows how the interceptor commit time varies with the intercept time (both measured before impact). The green curve in the plot is a construction curve that simplifies determination of the number of independent shot opportunities available. In these calculations, we assume there is sensor coverage of the entire trajectory. The first intercept we consider is launched at –1883 seconds and the intercept occurs at –814 seconds. If the intercept is successful, we are done. However, if the first intercept fails, there is still time for a second shot. If we wait 100 seconds and then launch the backup interceptor at –714 seconds, the intercept will occur at –180 seconds. At this point, there is no time for a third intercept. The number of shoot-look-shoot opportunities is called the depth of fire. If the depth of fire were three, we would say that we had SLSLS capability. Later, we will see that SLS offers the potential to reduce the system leakage at a significant savings in interceptors. However, to achieve SLS capability, we may need fast, long-range interceptors, sensor coverage over most of the trajectory, and an ability to assess whether the first intercept was successful or not. Again, we are faced with a quantity-quality trade-off.

Next we turn to questions of the sensor coverage needed to support the interceptor coverage available. Figure 10.5 illustrates why we maximize the coverage against long-range missiles by using uprange sensors to provide targeting for interceptors closer to the defended region.

If we have a collocated sensor and interceptor in the defended region, the sensor will not detect the threat until late in its trajectory. The detection time is limited by when the target is both above the local horizon and within the sensor's range. For an ICBM, detection time is typically in the last third of the trajectory. The defense footprint is limited by how far the interceptor can fly in the time between when it can be launched and when the intercept occurs. This distance is indicated by the green coverage region in the figure. However, if we had sensors near the missile launch point that could determine its trajectory much earlier, we would have nearly three times the time available for interceptor flyout, giving a much larger footprint. This coverage is indicated by the pink region in Figure 10.5. (To be more accurate in calculating the interceptor footprint, we look at trajectories to all points in the defended region. This process is more tedious, but it gives results similar to those obtained by using distance = speed × time.) In practice, having three times the time for interceptor flyout does not necessarily translate into three times the footprint radius since the interceptor may

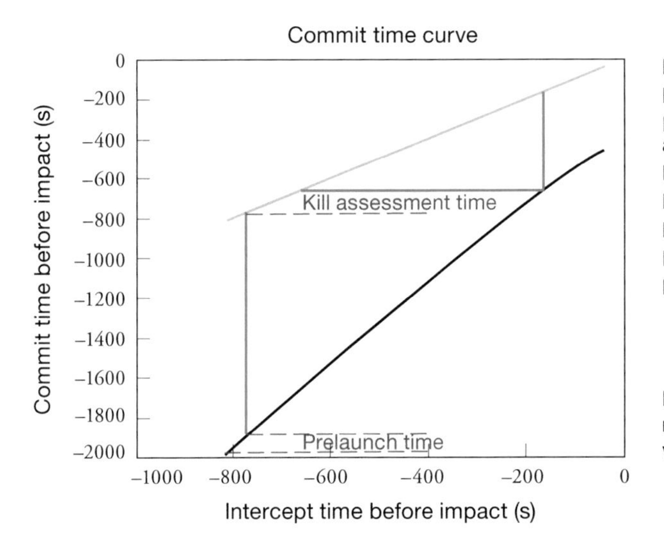

Burnout velocity	7 km/s
Delay time	100 s
Min. interceptor altitude	100 km
Reentry angle	25°
Prelaunch time	100 s
Kill assess time	100 s
Min. flyout time	200 s
Depth of fire	2

Kill assessment requires that a sensor view after the intercept

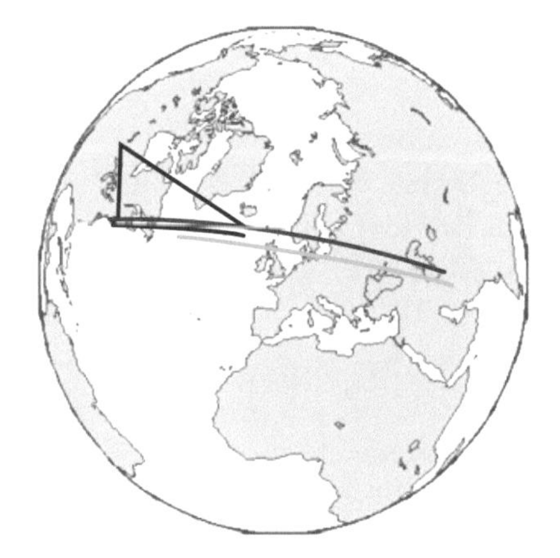

	Latitude°	Longitude°
Launch	36	56
Impact	42	–71
Interceptor	48	–97

Event time	Earliest (s)	Latest (s)
Launch	–1994	
Commit	–1883	–462
Intercept	–814	–38
Impact		0

Trajectory
Intercept window
Commit window

Figure 10.4 Interceptor timeline.

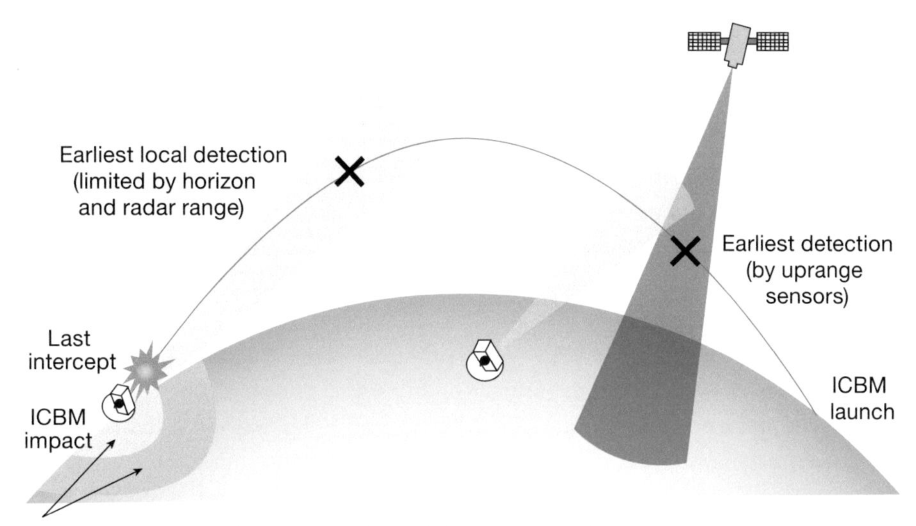

Figure 10.5 Determining the defense footprint.

not be able to fly that far no matter how much time it has. However, to get nationwide coverage from a single interceptor site, we need both a fast interceptor and sensors that can see the threat soon after it is launched. To achieve this greater coverage, it is necessary to launch the interceptor on the basis of uprange sensor data long before the local sensor can see the target. It took a while for system operators to trust the data from these distant sensors, but they really had no choice if they wanted national coverage. We now turn to sensor coverage.

10.3.2 Sensor Coverage

This section discusses what factors determine the range of a radar or infrared (IR) sensor. The range of a radar can be determined from the radar equation, which relates the radar performance to a few radar design parameters. The radar equation is derived in Appendix 10A; here we look at some of its applications. The form of the radar equation we use is

$$P A / kT L = 4 \pi \Omega R^4 (S/N) / \sigma t. \tag{10.3}$$

In this equation, P is the radar average power, A is the antenna area, Ω is the solid angle the radar has to search in time t, R is the radar range, S/N is the signal-to-noise ratio, σ is the target radar cross section, T is the receiver noise temperature, L is the total radar loss, and $k = 1.38 \times 10^{-23}$ (in metric units) is Boltzmann's constant. The fact that Boltzmann's constant is so small is what makes long-range radar possible. (The search solid angle Ω is a measure of the fraction of a sphere that we must search. If we searched the entire sphere, Ω would be 4π.) In Equation (10.3), we have rearranged the radar equation so that design parameters are on the left and performance parameters are on the right.

Making the right-hand side larger makes the radar more capable; it could search a larger sector faster at longer range and/or higher signal-to-noise ratio against smaller targets. Making the left-hand side larger makes the radar more expensive; it needs more power, larger antennas, and/or lower noise and losses. As in the case with interceptors, we have the option of deploying more lower-performance radars or fewer higher-performance radars: yet another trade-off between quantity and quality.

When I first learned about the radar equation, I was told that it is simple enough that anyone can learn to use it. I was also told that the radar equation is complicated enough that anyone can screw it up. It is essential to do a sanity check (as described in Section 10.5.6) before continuing with your analysis, and certainly before presenting your results.

One interesting aspect of the radar equation is that all the important factors enter multiplicatively. If we plot things on log-log paper, we can look at the interaction of six variables on one graph. Figure 10.6 shows such a graph for a search radar (I have become locally renowned at Lincoln Laboratory for stuffing everything onto one chart).

Here we plot the search range as a function of the radar average power and the antenna area. We have different scales for power depending on the search frame time, and we have different scales for aperture depending on the target radar cross section. We also plot different ranges for different search sectors in different colors. We could probably add another dimension by using dashed and dotted lines, but that might make the graph too hard to use (just kidding). As an example of the use of Figure 10.6, consider a radar with 100 kW of average power, and aperture of 100 m^2 searching a steradian in 10 seconds looking for a 0.1 m^2 target (assuming the T, L, and S/N values given in the figure). By working on the appropriate P and A scales, we find a maximum range of about 700 km using the black curves.

We can do a comparable, but less straightforward, analysis for IR sensors. These sensors are passive and respond to radiation emitted by the target and to radiation emitted by background sources, both internal and external to the sensor. For confident detection and track, the signal from the target needs to significantly exceed the noise from the background. Both signal and background depend on the temperature of the objects emitting the IR radiation. For detection of room-temperature targets, the sensors need to have cooled optics and focal planes, and to view the targets against a space background. Here we do not develop exact equations for the IR signal-to-noise ratio but indicate how the ratio scales with parameters such as search solid angle, search time, range, sensor aperture, focal plane size, and background and signal levels.

To derive an *optics equation*, it is useful to consider a number of resolution or field-of-view (FOV) variables in angular space. These variables are indicated in Figure 10.7. The actual derivation of the optics equation is given in Appendix 10B.

The form of the optics equation we use is

$$Q\,A\,/L\,I_b \sim (S/N)^2\,\Omega\,R^4\,/\,t\,I_t^2 \qquad (10.4)$$

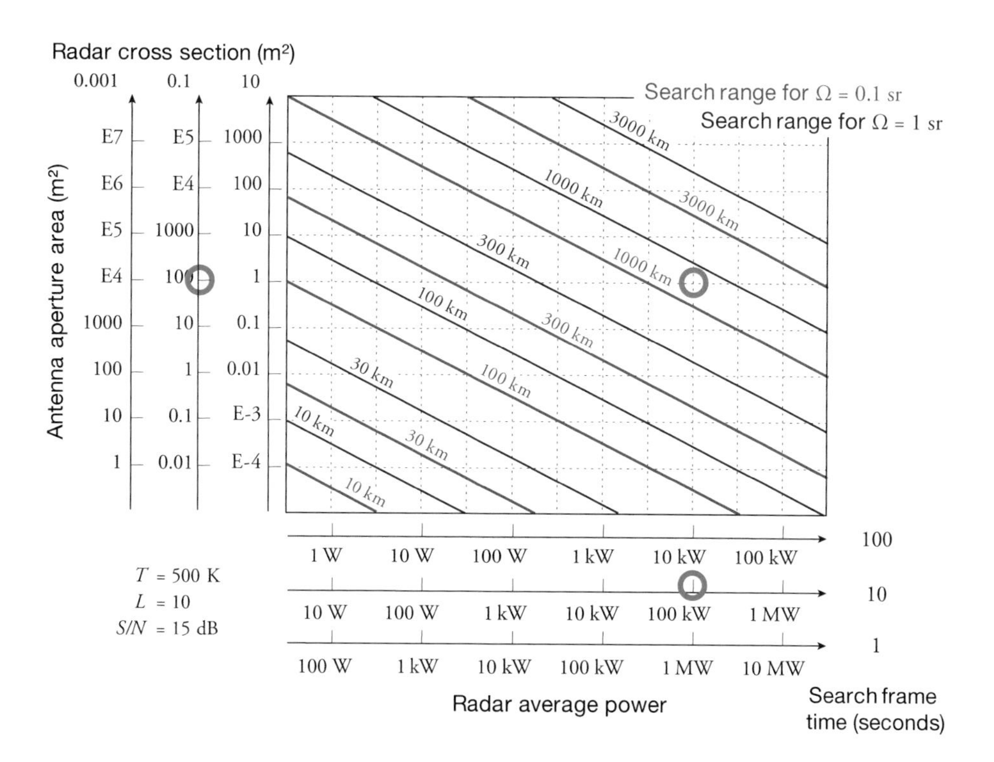

Figure 10.6 Scaling of the radar search equation.

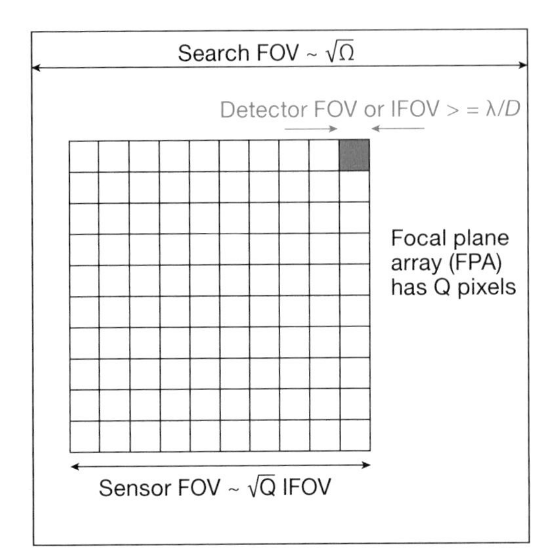

To search the solid angle, we need to move the sensor FOV to Ω/FOV^2 different positions

The dwell time in each position will be the total search time divided by the number of positions (assuming negligible scan and settle times)

Figure 10.7 Field-of-view (FOV) variables in angular space.

As with the radar equation, we have put design parameters on the left and performance parameters on the right. In Equation (10.4), A is the telescope area, Q is the number of pixels in the focal plane, I_b is the background intensity, and I_t is the target intensity. Ω, R, t, L, and S/N have the same meanings as in the radar equation.

In this form of the equation, on the left-hand (design) side, $Q\,A$ plays the role of the power-aperture product and I_b plays the role of the noise. Again, increasing the left side makes the sensor more expensive. On the right (performance), the dependence on Ω, t, and R is the same as in the radar equation. However, the dependence on the target intensity I_t and S/N is quadratic rather than linear as in the radar equation. This difference is not surprising since the sensor types are very different. More surprising are the similarities.

We now turn to the question of modeling the leakage of a BMD system. We make use of many of the coverage results obtained above in the following analysis.

10.4 Leakage Analysis

This section describes the major functions that a BMD system must perform: detection, identification, and kill. The performance of each of these functions depends on a number of properties of the threat objects and the defense sensors and interceptors. We can always improve the performance of a function by spending more money on the defense components or by buying more components. Detection can be limited by sensor coverage, as addressed above, but it can also be limited by not having enough sensitivity to overcome noise or unwanted background signals. Sensors are also affected by stealthy or fluctuating targets. In this section, we derive the detection leakage for a variety of cases, but the important result is that improving performance usually costs more. Identification performance depends on measuring differences between warheads and decoys. In this evolving offense versus defense battle, the losing side will respond with new countermeasures or counter-countermeasures. Here we do not look at specific targets or techniques; rather, we present a statistical model that tells us how good our identification has to be to avoid firing too many interceptors at false targets. Finally, we model the kill function by seeing how we need to continually improve our estimate of the target's trajectory and maneuver our interceptor to always keep the target within its coverage. The sections are essentially independent, so you can choose to read through only those topics that interest you.

Leakage is a measure of the quality of the defense. If the defense were perfect, the leakage would be zero. If the defense were worthless, the leakage would be one. Here we look at all the ways the defense can fail and try to understand the most important factors contributing to leakage. Since leakage is a probabilistic quantity, many of the calculations are of a probabilistic nature.

To appreciate how leakage happens in a BMD system, we need to consider the sequence of functions that must be done successfully. This sequence is often expressed in terms of a kill chain as shown in Figure 10.8.

In this kill chain, we have included two functions related to identification and two functions related to kill. The track function determines where the target is and where it is going, and the identification function determines what the target is. The attack function determines whether the interceptor can get to the target in time, and the kill function determines whether the interceptor can destroy the target. In this section, we simplify the kill chain to include just detect, identify, and kill (DIK), as indicated in Equation (10.2). We look at some of the things that can go wrong with

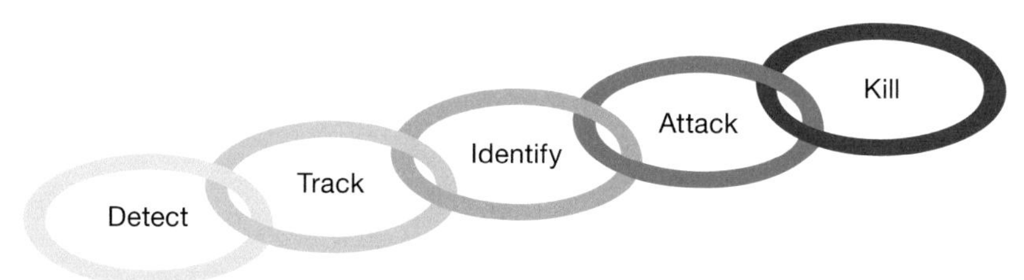

Figure 10.8 The BMD kill chain.

each function and thus produce leakage. We also look at what the defense can do to reduce this leakage. Again, we come up with quantity-quality trades to reduce leakage. Before getting into the sources of leakage, it is of interest to look at the consequences of leakage.

Figure 10.9 shows how the survival probability of a defended target depends on the leakage achieved by the BMD system and the size of the attack. Since it takes only one leak to destroy the target (assuming a nuclear-armed attacker), the target survives only if every attacker is killed. The formula for this is

$$\text{Target Survival Probability} = (1 - \text{Leakage})^{\text{Number of attackers}} \tag{10.5}$$

In making leakage calculations, we make use of the fact that the success probability is 1 minus the leakage probability. We use this relationship to simplify the equations and calculations.

The survival probability in Figure 10.9 is plotted linearly in the left plot and logarithmically in the right plot. From the left plot, we see the survival probability dropping dramatically as the leakage increases and as the attack size increases. From the right plot, we see how low the leakage has to be to achieve a high target survival probability. We now look at some of the factors contributing to leakage for the major defense functions of detection, identification, and kill. Detection and identification are done primarily by the sensors, while kill is done primarily by the interceptors.

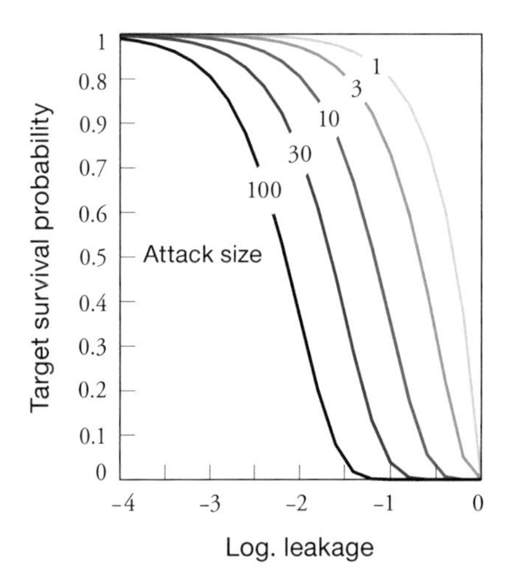

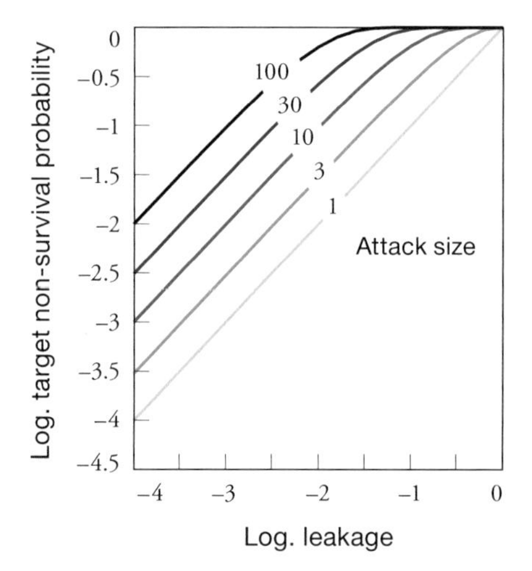

Figure 10.9 Consequences of leakage.

10.4.1 Detection

If an attacker is not detected or is detected too late, it does not matter how well we could perform identification and kill, the target will leak through the defense.

The leakage for the detection process can be determined through a combination of geography, geometry, and probability. In Section 10.2, we looked at the timeline needed to conduct a defensive engagement. This timeline presumed that target detection could be accomplished by a given time for each potential attacking trajectory. If detection is not achieved by this drop-dead time, leakage will result. The reasons why detection may fail include the following:

- Sensor is not close enough to threat
- Threat is below the horizon to the sensor
- Sensor is looking in the wrong direction
- Sensor does not have enough sensitivity to see target
- Too many targets are present for the sensor to handle
- Target is dimmer than expected
- Noise is higher than expected

Some of these failure modes are deterministic while others are probabilistic. The geography and geometry considerations usually result in either 0% or 100% leakage while the sensitivity and signal-to-noise considerations result in a probabilistic leakage. An example of issues with geography and geometry is shown in Figure 10.10. Here we look at two trajectories from the Middle East to the United States as viewed by a single-face array radar in the eastern Mediterranean Sea. This radar can only see targets that are above the horizon and within a 90-degree angular sector. This is the radar field of view (FOV).

For the red trajectory to the East Coast, the threat is within the radar's field of view for a long while. For the trajectory in blue impacting in Alaska, the threat never gets above the radar horizon. The radar will not detect targets that are outside its FOV no matter how capable it is. Additional sensor sites would be needed to detect these threats. For targets that get within the radar's FOV, the detection probability depends on the radar sensitivity, the target range, and the target radar cross section. Calculation of the detection probability for those cases is indicated below.

For those targets outside the radar's FOV, we have a number of options if we cannot accept this leakage. We can deploy additional radars to cover some of the gaps in detection coverage. This is where geography comes into play. To get early coverage of

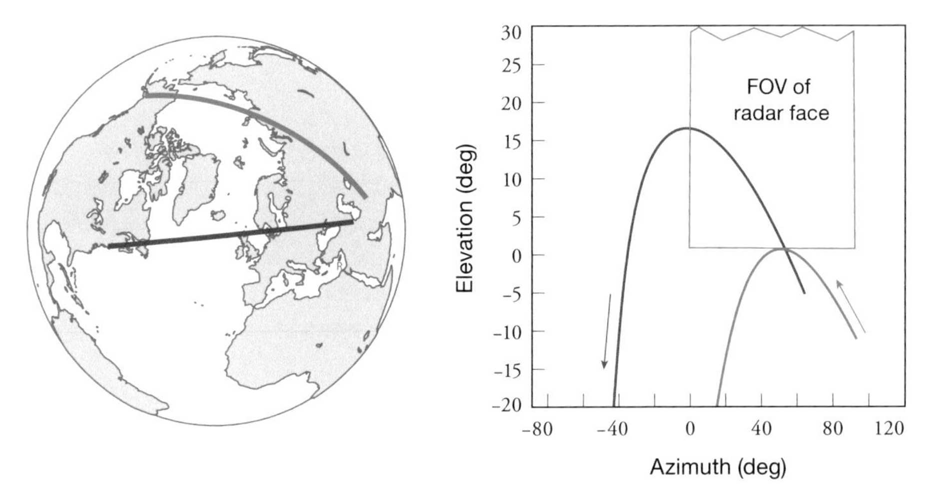

Figure 10.10 Radar visibility for different trajectories.

all trajectories to Alaska, it would be necessary to have sensors deployed in the former Soviet Union or on aircraft or in space. An alternative would be to wait until the trajectory becomes visible to sensors in Alaska or the Arctic Ocean. In this case, we might need more interceptor sites or faster interceptors to make up for the time we lost by not detecting earlier. Very often, it is necessary to deploy a number of defense sensors to provide the required detection coverage for all threatening trajectories. Since we do not know in advance which of the sensors will detect the threat, it is necessary to net the sensors together so that any one of them could provide information needed to launch interceptors. This arrangement is particularly necessary for systems that are responsible for a large defended area and/or against a large attacking country. For defense of small high-value targets, a point defense with collocated sensors and interceptors would be adequate.

We now turn to the case in which the trajectory passes through the sensor's FOV and look at how the detection probability depends on the target's signal-to-noise ratio and the detection approach the defense takes. Figure 10.11 indicates how the signal level compares with the noise level for a variety of cases. These plots are of probability density as a function of signal level. The sensor is most likely to see signal levels near the peaks of the curves, but there is a small probability that the sensor will see smaller signals when the target is present and larger signals when the target is absent. It is these unlikely events that result in detection leakage and false alarms.

We consider cases with and without noise and targets with constant and fluctuating signatures. In each case, there is a probability distribution of the signal level of the resolution cell with the target and also of all the cells with only noise. For detection, there may be millions of resolution cells in the search volume, and it is important that the probability of detecting noise be as small as possible. Traditionally, we set a detection threshold well above the mean noise level to reduce the false-alarm rate (P_{fa}) to 10^{-6} or 10^{-8} so that there are only a few false alarms in each search frame. The detection probability is then the probability that the target return is above this threshold. We can calculate the detection probability as a function of S/N level, and sample results for a radar sensor are shown in Figure 10.12 for different false-alarm rates.

Note that the S/N scale is in decibels (dB). This is a logarithmic scale where a difference of 10 dB is a factor of 10. Thus 0 dB = 1, 10 dB = 10, 20 dB = 100, 30 dB = 1000, etc. We see that the detection probability (P_d) increases steadily with increasing S/N as expected. If we require a lower probability of false alarm, the detection probability decreases for the same S/N. We also see that the curves for the fluctuating target have more gradual slopes than for the constant target. At low S/N levels, the fluctuation of the target may cause the measured signal to occasionally exceed the detection threshold, while for the constant target, only noise could cause this. However, for high S/N levels, the constant target has a significantly higher detection probability than the fluctuating target. In this region, target fluctuations may cause the measured signal to fall below the detection threshold. At the very highest detection probabilities, we can

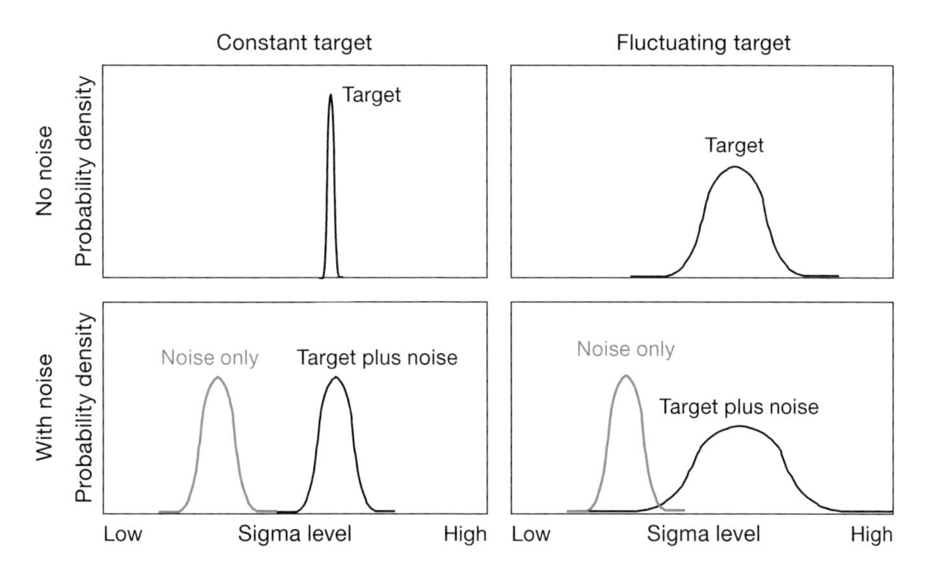

Figure 10.11 Signal-level detection statistics.

use different strategies to detect fluctuating targets and save significant radar energy. We note that the solid red curve indicates we can achieve a P_d of 0.9 at a S/N of 17 dB, but to achieve a P_d of 0.99, we need about 23 dB S/N. If, instead of using an extra 6 dB of energy, we use two independent pulses, each with 17 dB S/N, and accept a detection on either pulse, we have a probability of 0.1 of failing to detect on each pulse and a probability of 0.01 of failing to detect on both pulses. Since a factor of 2 is 3 dB, we use only 20 dB worth of energy to get essentially the same $P_d = 0.99$. In more severe target fluctuation models, this payoff is even greater.

In the above discussion, we see that detection probability increases with signal-to-noise level. In the previous section's Equation (10.3), we saw that, for a given radar, the detection range will decrease if we require increased signal to noise. With reduced detection range, we will have reduced coverage. In addition to quantity-quality trades, we also have coverage-leakage trades. In fact, leakage varies over the coverage area, so we have these trades at all points in the defended region.

The above analysis has been for a radar sensor, but similar results can be obtained for IR and optical sensors. The basic results are that the P_d increases with S/N and P_{fa} and that target fluctuations make it more expensive to achieve a very high P_d. We now turn to the next step in the kill chain, target identification.

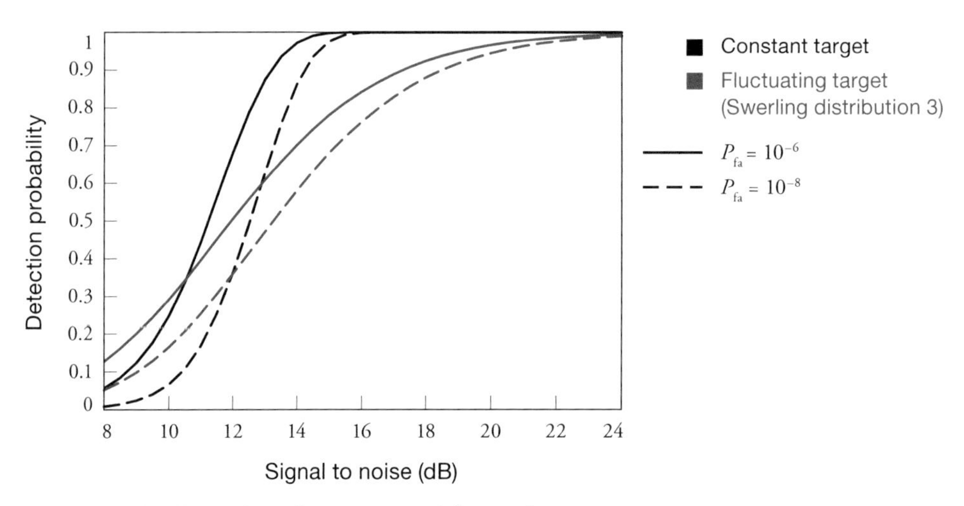

Figure 10.12 Detection of constant and fluctuating targets.

10.4.2 Identification

The process of target identification consists of answering questions: What is the target? Where is the target? Where is the target going? The "what" question is usually called discrimination and the "where" questions are usually called tracking. The tracking questions are relatively easy to analyze. Each measurement on a target tells us something about its position in space and possibly its velocity. By combining a sequence of measurements, we can improve our estimate of the target state vector consisting of both its position and velocity vectors at any instant of time. When the target is outside the atmosphere, it follows a predictable ballistic trajectory so that the state vector at any time determines the state vector at future times. Associated with the state vector is a covariance matrix, which determines our uncertainty in the target state. When we predict the future trajectory, uncertainty in velocity translates into future uncertainty in position. The size of the uncertainty volume is a function of the sensor measurement accuracy, the track time, and the measurement rate. The sensor accuracy is a combination of the sensor bias (it isn't pointing where we think it is pointing) and the sensor precision. If multiple sensors track a target, the uncertainty volume can be significantly reduced from that for a single sensor, provided that the measurements for the same target are correctly associated. This association problem can be very difficult if there are large numbers of closely spaced targets viewed by the sensors.

In general, radar sensors tend to measure range very accurately and angles (azimuth and elevation) relatively poorly. The uncertainty volume for a radar tends to be a thin pancake. Passive optical sensors are just the opposite. They measure angles fairly accurately and do not measure range at all. A passive sensor can infer range by tracking for a long time and observing the curvature of the trajectory in the gravitational field. Conversely, we can determine range by using two passive sensors in a triangulation or stereo mode. Again, this configuration works only if the two sensors correctly associate the targets.

For targets in the boost or atmospheric reentry phases, tracking and prediction are much more uncertain because of unknown target accelerations. Uncertainties in acceleration very quickly result in large uncertainties in position. In many cases, the target acceleration is parallel to the velocity vector. This is the case for boosting targets where the thrust is along the velocity vector and also for targets in reentry where the drag is along the velocity vector. In these cases, we can write the position uncertainty at a future time $\sigma_x(t)$ in terms of the position, velocity, and acceleration uncertainties as

$$\sigma_x(t)^2 = \sigma_x(0)^2 + \sigma_v(0)^2\, t^2 + (\sigma_{acc}(0)/2)^2\, t^4\,. \tag{10.6}$$

This equation is graphed in Figure 10.13 for some sample values on both linear and log paper. We see that, at different times, different error sources dominate the prediction error. If we observe some portion of the missiles' boost phase but do not observe the booster burnout (the start of the ballistic phase of the trajectory) and try to predict the target position well into midcourse flight, the predicted position error could easily be hundreds or even thousands of kilometers dominated by the acceleration uncertainty. If we observe the missile after burnout, the acceleration uncertainty is zero and the velocity uncertainty is usually the dominant error source.

These prediction errors can translate into interceptor leakage for a number of reasons. For a single target, if the prediction error is larger than the interceptor's divert capability, there is a reduced probability that the interceptor will be able to home on and hit the target. If there are multiple targets in or near the predicted intercept position, it is possible that the interceptor will select the wrong target, also contributing to leakage. These leakage sources will be discussed further in the next subsection. Now we turn to the possibility of discrimination leakage, which results when the sensors identify the wrong target as the warhead and do not attack the real warhead.

The modeling of the discrimination process is similar to that of the detection process discussed above. Here we are trying to distinguish between a warhead and a decoy on

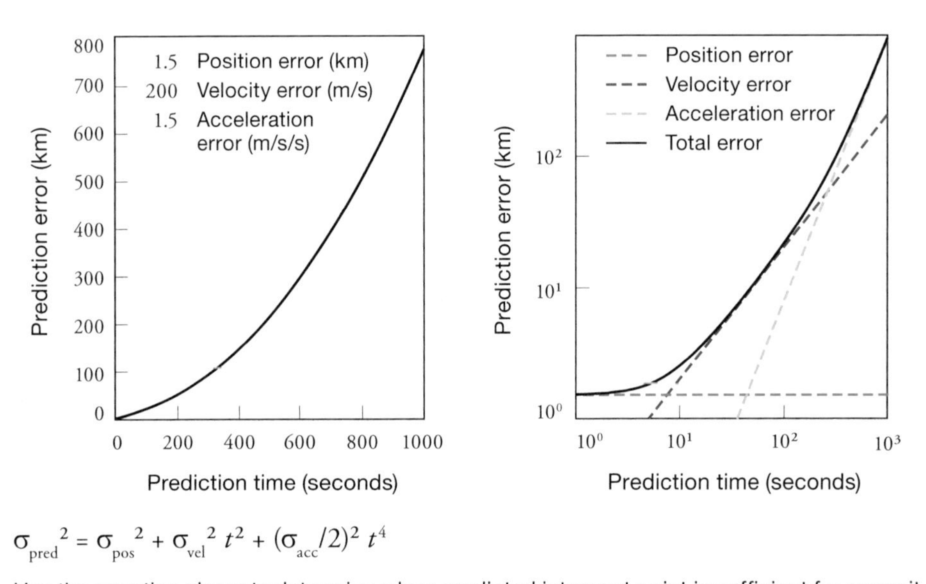

$$\sigma_{pred}^2 = \sigma_{pos}^2 + \sigma_{vel}^2\, t^2 + (\sigma_{acc}/2)^2\, t^4$$

Use the equation above to determine when predicted intercept point is sufficient for commit

Figure 10.13 Radar visibility for different trajectories.

the basis of sensor measurements. This case is similar to that shown in Figure 10.11 in that both decoys and warheads have probability distributions in whatever dimension can be used to distinguish them. However, decoys differ from noise in that they are deliberately designed to look like warheads. The discrimination problem also differs from the detection problem in that we are trying to reject small numbers of decoys (from a few to a few hundred) rather than the millions of resolution cells containing only noise.

The ability of the defense to find dimensions that separate decoys and warheads depends on the design of the decoys and warheads, the defense intelligence information on these designs, and also the measurement capability of the defense sensors. When in ballistic flight outside the atmosphere, a lightweight decoy will follow essentially the same trajectory as a heavy warhead. Without my getting into specific details, the ability to discriminate depends on the signatures of the warheads and decoys as a function of time, space, and wavelength. The time dependence of the signature tells something about the angular motion of the target. The space dependence tells something about the size and shape of the target. For radars, the wavelength dependence tells something about the shape and surface material of the target. For IR wavelengths, the wavelength dependence tells something about the temperature and surface material of the target. In general, if we have measured data on potential threat objects, we can develop pattern recognition algorithms that could discriminate the objects in a real attack. However, if there are threat objects we have never observed, we need more robust discrimination techniques. In what follows, we will not discuss specific discrimination techniques but will describe methods of modeling discrimination performance and determining discrimination requirements for different scenarios.

Figure 10.14 describes a very useful simple model of discrimination performance and is patterned after the detection model described above. We model both the warhead (we call it RV for reentry vehicle) and decoy distributions as Gaussian with equal standard deviations in a discrimination dimension. The separation of the means of the distributions is k times this common standard deviation. We have thus reduced the discrimination performance to a single number. A high k-factor is indicative of good defense (good discrimination and low leakage). In an attack, the defense sets a threshold and classifies all objects exceeding this threshold as warheads and all other objects as decoys. The defense makes a mistake if it calls a decoy a warhead (false alarm) or if it calls a warhead a decoy (leakage). Later, we see how this threshold can be set depending on the number of decoys, the number of defense interceptors, and the tolerable discrimination leakage. The trade-off among leakage, false alarms, and the k-factor is shown in the three contour plots in Figure 10.14. Each pair of leakage (P_L) and false-alarm (P_{fa}) probabilities corresponds to a specific k-factor. Conversely, each value of k corresponds to a specific trade-off between leakage and false alarms. If leakage and false-alarm rates are plotted on probability paper, the lines of constant k are straight and equally spaced. They are more complicated on linear or log paper. Note that we show plots for negative values of k. These values correspond to cases in which the decoys look more like warheads than the warheads do. This confusion could result if the defense had incorrect intelligence information on what the threat objects look like.

We now look at how we can compensate for discrimination leakage by using additional interceptors. Figure 10.15 shows how we can calculate the resulting leakage.

Here we assume an attack of one RV and 10 decoys. Depending on the discrimination k-factor, we will operate on one of the curves. If we have N interceptors to use on this attack, we are constrained to operate on one of the ascending straight lines.

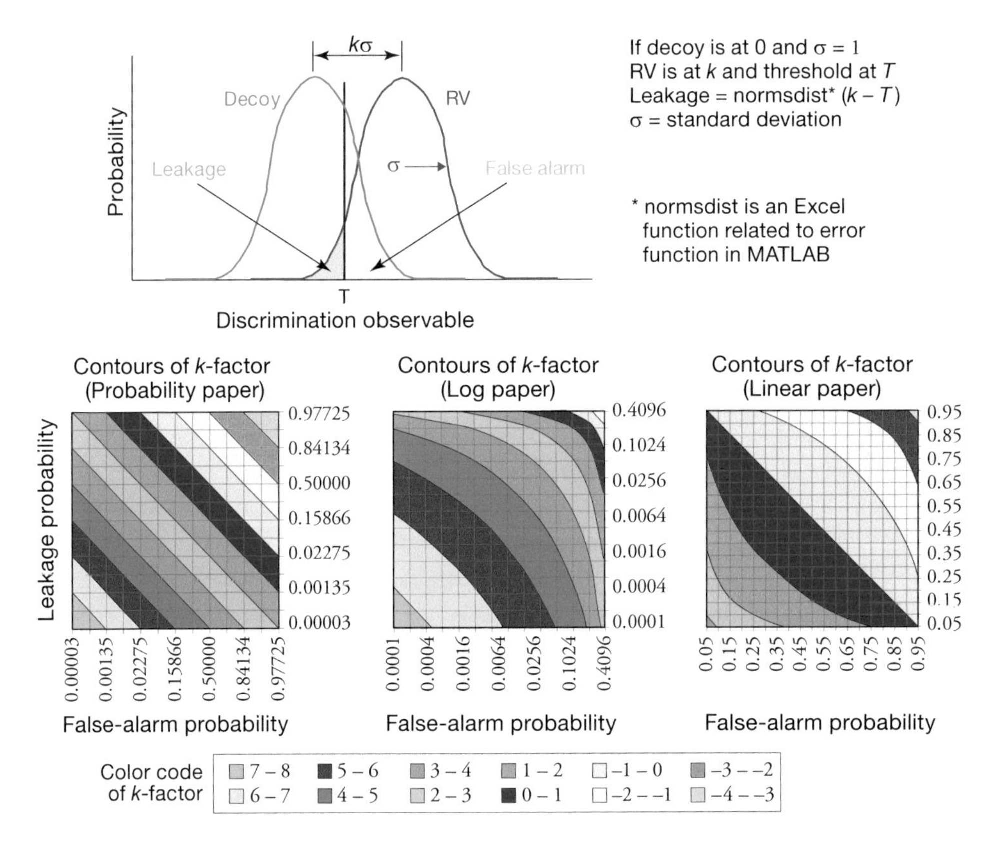

Figure 10.14. The k-factor model.

These lines correspond to the cases that will use one, two, or three interceptors. Where the interceptor curve crosses the *k*-factor curve is our operating point, and we can read off the resulting discrimination leakage directly—this is the probability that we will classify the RV as a decoy and not shoot at it. We see that we can reduce leakage either by improving the *k*-factor or by using more interceptors. Figure 10.16 shows this trade-off for a wider range of parameters.

We see that, for high leakage, we can compensate for a low *k*-factor with a small number of additional interceptors, but to achieve low leakage, we really need a higher *k*-factor. Later in this section, we will see how we can make an integrated treatment of the leakage resulting from discrimination and the leakage resulting from interceptor unreliability.

One final point to note in this subsection is the dependence of discrimination performance on the number of decoys in the attack. To appreciate this trade, we need to look at things from the attacker's point of view. Figure 10.17 looks at some of the attacker's options.

The offense does not get decoys for free. If the decoys do not draw interceptors, they are a waste and the attacker should not use them. Figure 10.17 illustrates some of these points. For ease of explanation, the figure illustrates a case of a missile that can

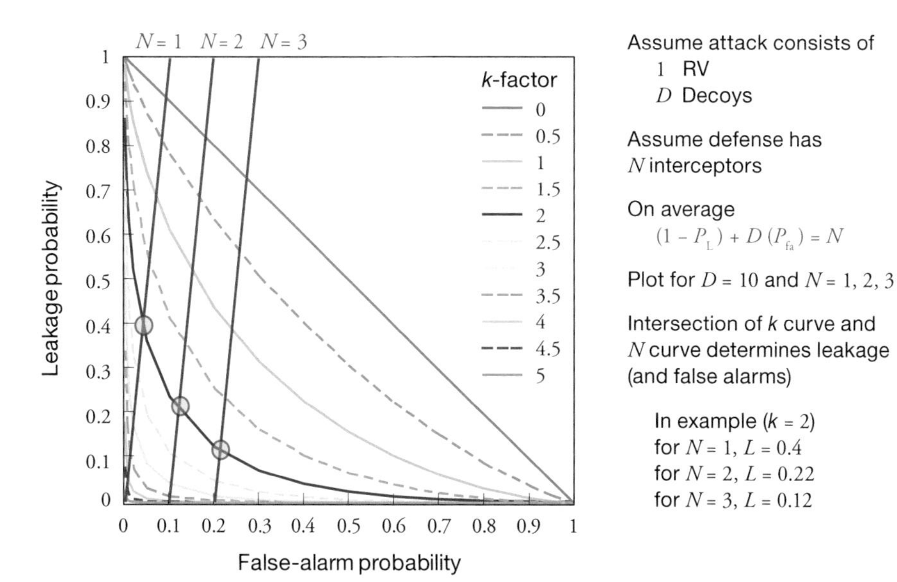

Figure 10.15 Trading interceptors for discrimination.

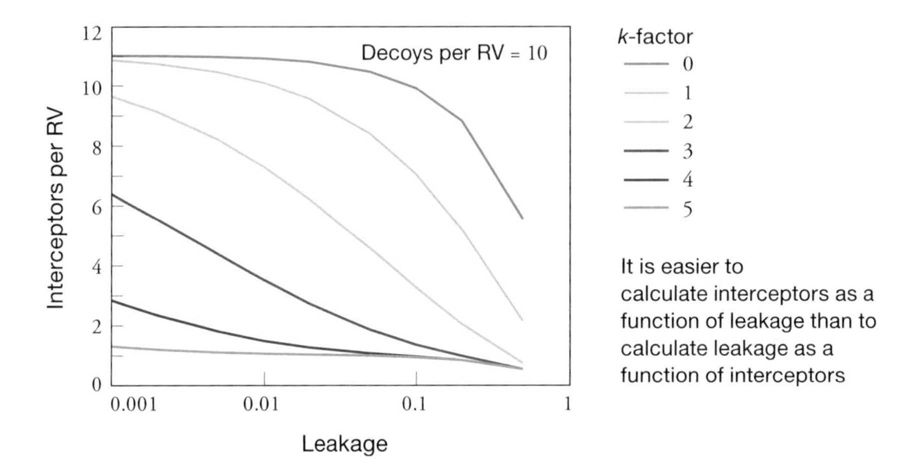

Figure 10.16 Trading interceptors for discrimination with a wider range of parameters.

carry 10 RVs. If the offense wants to use decoys, it must offload some of the RVs. We look at cases in which the offense can switch one RV for 5, 10, 20, 50, or 100 decoys. The offense can choose what type of decoy to use and how many RVs to replace with decoys, and can operate on any of the curves in the figure. Clearly, the offense does not want to offload all the RVs because decoys don't kill anything. The lighter decoys should be easier to discriminate but may be so numerous that they overload the defense sensors; these are called traffic decoys. The heavier decoys are harder to discriminate, but the defense may be able to afford to shoot at all of them. The attacker must assess these possibilities and select a specific threat.

In the "Goldilocks" comparison in Figure 10.17, the offense has selected an intermediate-sized decoy as its best choice. The defense might be tempted to assume this decoy will be the actual threat and tailor its design to defeat this threat. In particular, the defense might develop a discriminant that can reject these intermediate decoys and just shoot one interceptor at the RV. Alternately, the defense may not discriminate at all and shoot at the RV and the four decoys. Either of these strategies works if the offense uses its optimum strategy. However, they will both fail if the offense uses one of its suboptimum strategies. If the defense had a discriminant that could reject intermediate decoys but not heavy decoys, it would have to use two interceptors (as opposed to one for the optimum attack) against the left attack of one RV and one heavy decoy. Conversely, if the defense just shot at all the decoys and the offense used the middle attack with 15 traffic decoys, the defense would have to use 16 interceptors (as opposed to 5 for the optimum attack). To handle whatever the offense throws at it,

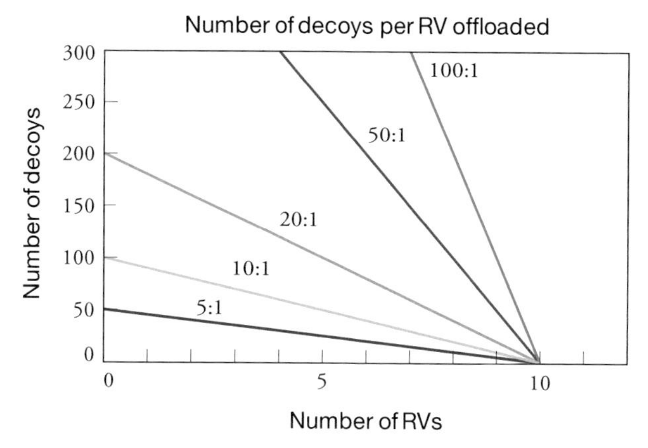

This is a typical example of a common "Goldilocks" analysis

Figure 10.17 Attacker's payload options.

the defense needs enough interceptors to shoot at the decoys it can't discriminate and enough discrimination to reject decoys it can't afford to shoot at. These requirements result in a more expensive (but more robust) defense.

We can determine what kind of *k*-factor is needed to eliminate the offense's payoff from using decoys by looking at Figure 10.17's plot in more detail. We can go through a sample calculation to indicate the process. Suppose the offense uses 40:1 decoys and offloads two RVs. The attack then consists of 80 decoys and eight RVs, which is the 10-decoys-per-RV ratio used in Figures 10.15 and 10.16. If the interceptors were perfect, we would need 10 interceptors to handle the threat with all RVs. With only eight RVs, we can afford to shoot at two of the 80 decoys for a false-alarm rate of 0.025; this is an interceptor-to-RV ratio of 1.25. As shown in Figure 10.16, we can achieve leakages of 1% or below with a *k*-factor of about 4.5 or higher in this scenario. Repeating the analysis with 10:1 decoys and offloading five RVs, we get the same 10-decoys-per-RV ratio (five RVs and 50 decoys). However, with our 10 interceptors, we can afford an interceptor-to-RV ratio of 2:1. In this case, we could achieve leakages of 1% or below with a *k*-factor of 3.7 or higher. It is not clear whether it is easier to get a *k*-factor of 3.7 on a 10:1 decoy or of 4.5 on a 50:1 decoy, but these numbers can provide guidelines for decoy designers and discriminators to see what kind of performance would make a difference. Here we have looked only at two very specific scenarios. To determine the most stressing cases for discrimination, we would have to examine the entire threat space in Figure 10.17 and also consider actual decoys and discriminants rather than just assuming *k*-values. However, to get a simple representation of BMD system performance, the *k*-factor is an excellent first step. In the next section, we look at how discrimination and intercept performance can be traded off to maximize overall system performance. First, we look at some important kill function issues and the leakage that can result if things go wrong.

10.4.3 Kill

To appreciate all the things that might go wrong in the intercept process, it is useful to think of the overall BMD engagement as consisting of a number of steps in which we refine our knowledge of where we want to send our interceptor. This direction to the interceptor is done by sensors off and on board the interceptor, each of which takes target information from previous sensors, refines it, and passes it on to subsequent sensors. Leakage can result from errors made by each sensor and also from failures of target association from sensor to sensor. As the information is refined, we have a better idea of where we want to send the interceptor. As the engagement progresses, the interceptor maneuvers to ensure that it can cover the remaining uncertainty in the final intercept point. As time goes by, we know where we want to go with better and better accuracy, but we have less and less time to divert to cover this uncertainty. Whenever the remaining uncertainty is larger than the interceptor's remaining divert coverage, leakage may result. Figure 10.18 illustrates this process.

Here we look at how the accuracy of our estimate of the future intercept point evolves as time passes and more sensors contribute to our knowledge of the target. We compare this with the ability of the interceptor to divert to cover this uncertainty volume. Note that accuracy is plotted on a log scale and varies over eight orders of magnitude. The two curves correspond to the uncertainty in our estimate of the intercept point (green) and the remaining divert capability of the interceptor (blue). These curves depend on the particular trajectory and the location and capability of the sensors and interceptors; the curves in the figure should be regarded as examples rather than analysis results. We can start from the right and see what happens as the trajectory evolves. Prior to threat launch, we have a very large uncertainty about where we will make an intercept, corresponding to the size of our defended area constrained by the range of the attacking missile. We take this to be 10,000 km. Our interceptor has

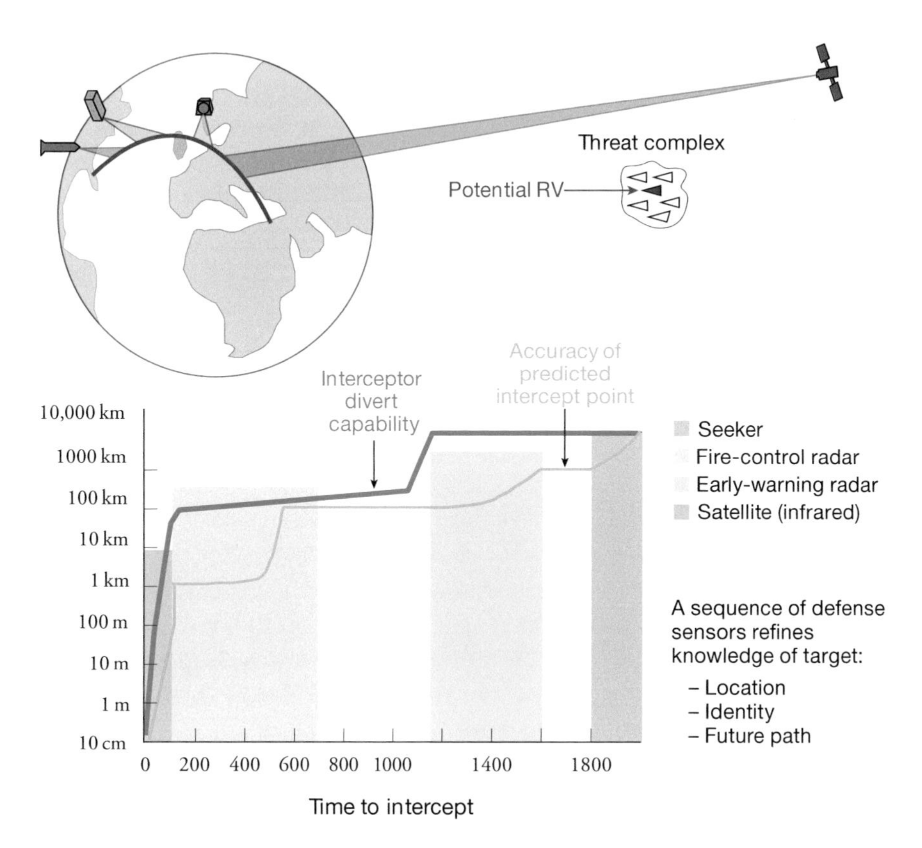

Figure 10.18 BMD scenario.

sufficient flyout capability to cover any trajectory to the defended area, and the blue curve is above the green curve. Once the missile is launched, an IR satellite will track the booster and provide more information on where it is going. As the missile burns its fuel, its future trajectory becomes more and more limited. The satellite tracking system (using two satellites to determine the range that is unavailable from a single passive sensor) measures the missile position as a function of time as long as its exhaust plume is bright enough to see. This measurement is represented by the pink region in the figure. After the final rocket stage burns out, the missile is on a ballistic trajectory that is deterministic. The satellite system tracks at a fairly low data rate and may not determine the exact burnout time, resulting in a burnout velocity uncertainty, which is the target acceleration times the burnout time uncertainty. For the case shown, this velocity uncertainty may result in an intercept point uncertainty of up to 1000 km. Since there is plenty of time left, we do not yet have to launch the interceptor, and the blue curve remains well above the green curve. We now enter a white region where there are no sensor measurements. It is likely that the missile will deploy various targets including RVs and decoys with varying separation velocities, creating a threat complex that may grow to 100 km in radius. Because no measurements are made in this region, there is no improvement in our estimate of the intercept point.

Later, the threat enters the coverage of an early-warning radar (light blue region), and we get a big improvement in target location. Since the radar can measure range, it can quickly refine the error volume provided by the satellite sensors. However, since the radar may not be able to identify the RV in the threat cloud, the intercept point uncertainty will be comparable with the size of the cloud. As time goes by, we eventually reach the interceptor drop-dead time. If we do not launch the interceptor by this time, it will be unable to reach the threat. At or before this drop-dead time, we launch the interceptor toward the center of the threat uncertainty volume. Once the main stages of the interceptor have burnt out, the interceptor divert coverage is determined by the smaller total divert velocity, ΔV, that the kill vehicle (KV) can achieve. The residual divert distance decreases as the time to intercept decreases. If we did not get more information on the threat, the divert capability of the interceptor would drop below the intercept point uncertainty and the leakage would increase significantly. Fortunately, in this case, the threat comes within the coverage of the fire-control radar (shown in tan), and we get more information on the projected intercept point. This information comes in two important flavors. First, and most importantly, the fire-control radar performs discrimination and determines which of the targets in the threat complex we should intercept. This action is shown as a large drop in the intercept point uncertainty from the size of the threat complex to the prediction accuracy for the target to be intercepted. Typically, this discrimination process takes a certain length of time (about 100 seconds in the figure). In addition, the fire-control radar generally has much better track accuracy than the early-warning radar has. The net result is to reduce the intercept point uncertainty to about a kilometer, which is well within the remaining interceptor divert capability. At this point, the defense has several options. In the old days, defense interceptors carried nuclear warheads with lethal radii on the order of kilometers. For such a system, we already have sufficient accuracy, and the interceptor could just fly to the predicted intercept point and detonate. However, if we wish to use a non-nuclear warhead or even a hit-to-kill (HTK) intercept, we need to achieve three or four orders of magnitude better accuracy in the predicted intercept point. This accuracy is most easily achieved by relying on a homing interceptor with an IR seeker providing the needed accuracy. The seeker achieves its accuracy in two ways; by using IR as opposed to radar sensing, much better angle accuracies can be obtained, and as the distance from the seeker to the target decreases, the position accuracy (angle accuracy times range) also decreases. However, as the interceptor-to-target range decreases, the time available for diverting the interceptor also decreases. There is again a race between improved accuracy and decreased ability to make use of this accuracy. The point at which the blue and green curves in the figure cross gives us an estimate of the expected miss distance. If this distance is smaller than the size of the target and the interceptor, we will achieve HTK.

It is important to appreciate the statistical nature of many of the results discussed above. In general, the target uncertainty volume is an ellipsoid with greater probability density of the target being near the center of the ellipsoid and decreasing probability density toward the edges of the ellipsoid. When we project this ellipsoid into the plane perpendicular to the interceptor-target closing velocity vector, we get an ellipse as shown in Figure 10.19. The interceptor divert capability, in this plane, is a circle. The radius of the circle varies around a nominal value, but this variation is generally much less than the radius itself. For simplicity, we can take it as a constant radius. In this case, the leakage probability is the probability that the target position lies outside the interceptor divert radius. By rescaling the coordinates, we can make the target distribution circular and integrate over an elliptical region to get the leakage probability.

The resulting leakage probability is shown for different values of the major and minor axes of the target uncertainty volume normalized by the interceptor divert radius in Figure 10.20. The leakage probability is shown by color as defined in the color bar at the right of each plot.

The left plot is linear in leakage while the right plot shows contours of the log of the leakage; it is appropriate for very low leakage numbers. (The log of leakage going from 0 to –4 corresponds to the leakage going from 1 to 0.0001.) Figure 10.20 shows contours of leakage that results whenever the blue curve in Figure 10.18 (interceptor divert capability) approaches the green curve (uncertainty in intercept point). These near crossovers can occur just before a new sensor acquires the target and starts to contribute its information or just after the interceptor maneuvers and loses some of its divert capability. These leakage contributions are additive; if there is a 70% probability that the early-warning radar prediction is inside the interceptor divert just before the fire-control radar acquires and another 70% probability that the fire-control radar prediction is inside the interceptor divert just before the seeker acquires, then there will be, at most, a 49% (70% times 70%) probability of a successful intercept. The plots show that to achieve leakages of a few percent or less, we need to have both semi-axes of the handover volume smaller than about a third to a half of the interceptor divert radius.

Additional complications may occur when transferring a target from one sensor to another (i.e., radar to seeker). The uncertainty volumes of the two sensors may differ

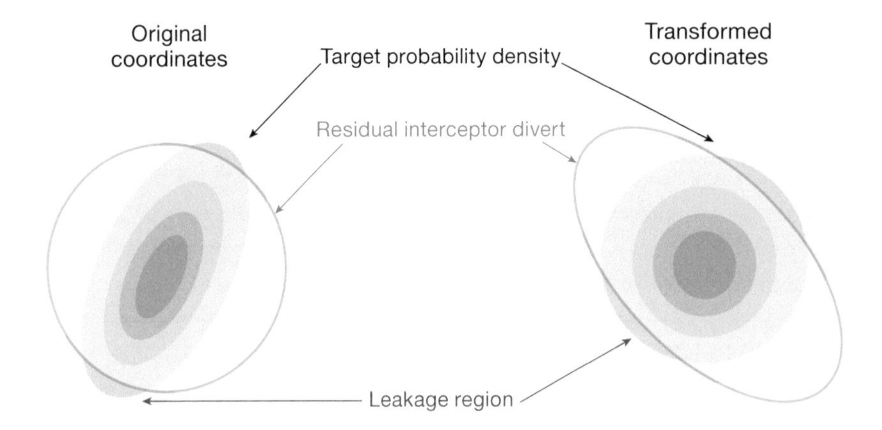

Figure 10.19 Leakage if interceptor cannot cover target uncertainty.

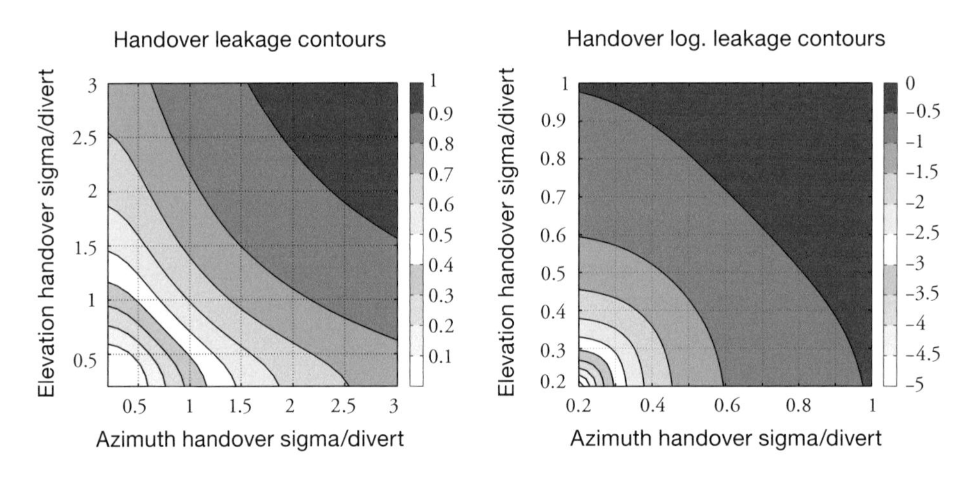

Figure 10.20 Handover leakage as a function of sensor accuracy and interceptor divert.

significantly in size, shape, and orientation. Multiple targets may be in one or both of the uncertainty volumes. A radar sensor may see targets or clutter that an IR sensor does not see, and vice versa. It is necessary to consider all these factors and possibilities when assessing the ability of the defense to conduct many of its required functions.

The final measure in the intercept process is the miss distance. Again, this is a race between the improving accuracy provided by the seeker and the decreasing divert capability of the interceptor as time runs out. To have a successful HTK intercept, we need the miss distance to be smaller than the target or interceptor size. Since the miss distance is a statistical quantity, we will have a certain probability of kill (P_k), which is another multiplicative factor in the overall engagement success probability. Often, we combine all the probabilities regarding the intercept process into an effective P_k.

Given all the things that can go wrong in the intercept process, it looks as if it will be very hard to achieve the low levels of leakage indicated in Figure 10.9. However, we have the option of firing multiple interceptors. If one interceptor has a probability of kill of 0.8, then the probability that at least one out of two interceptors will kill the target is 0.96. (The probability that one interceptor leaks is 0.2; the probability that both leak is 0.04.) If we have enough battlespace, it is possible to only launch the second interceptor after we have seen that the first interceptor failed. In this case, we expect to use only 1.2 interceptors on average to achieve the same 0.96 success probability. We only fire the backup interceptor in those 20% of cases in which the first interceptor failed. Going back to Figure 10.4, we see that the defense timeline needs to be sufficient to permit this shoot-look-shoot strategy. Often we can achieve SLS in the central portion of the defense coverage, but at the edges of the coverage, we have to salvo multiple interceptors to reduce the leakage to low levels.

In the next section, we look at strategies that try to balance the leakage from multiple sources (such as discrimination and interception), and additional strategies will be considered.

10.5 Balancing Leakage Sources

This section shows examples of some trade-offs among the different leakage sources. The main idea is that we can compensate for leakage by firing more interceptors. In particular, we can compensate for identification failures by shooting at some targets we think are decoys but are not sure about. We can compensate for unreliable interceptors by firing multiple interceptors at each target. The model presented here balances these two leakage sources to get the most efficient use of interceptors. This trade-off is typical of those conducted in BMD system design and analysis. It is not necessary to go through the mathematical details to appreciate that there are many ways to use our interceptors and that it is possible to find more efficient ways through systems analysis.

In the preceding section, we looked at leakage sources individually. If all the functions but one were perfect, the leakage from the imperfect function would be the total leakage. However, if several functions were imperfect, we could do better than just combining the leakage from each source. Here we give examples of how discrimination leakage and interceptor leakage can be reduced if the functions are treated together.

One way to minimize discrimination leakage is to fire interceptors at targets that we think, but are not quite sure, are decoys. One way to minimize intercept leakage is to fire additional interceptors at targets we think are RVs. If we treated these leakage sources independently, we might decide to shoot at the RV and two decoys to minimize discrimination leakage and to shoot two interceptors at each target to minimize intercept leakage. This scheme would require a total of six interceptors. If we treat the discrimination and intercept function together, we might fire two interceptors at the target we are pretty sure is an RV and one at each of the targets we think, but are not positive, are decoys. Here we will use four interceptors to achieve almost the same leakage that we had achieved in the case of firing six interceptors. If we know the interceptor P_k and the discrimination *k*-factor, we can find an optimal firing strategy to minimize the leakage for a given number of interceptors or minimize the number of interceptors needed to achieve a given leakage. Figure 10.21 shows how this process works for four interceptors against an attack of one RV and 10 decoys.

Figure 10.21 illustrates a procedure for deciding how many interceptors to fire at each threat object. We have multiple thresholds for the discrimination observable. If an object is clearly an RV, we will fire two interceptors at it. If it is an uncertain object, we may fire one interceptor at it. Finally, if it is clearly a decoy, we will not shoot at all. Figure 10.22 shows the overall leakage we can achieve for different numbers of interceptors as a result of this optimization process.

This plot is analogous to the plot in Figure 10.16, which assumed perfect interceptors. When we have both imperfect discrimination and imperfect interceptors, we save quite a bit over treating the leakage sources independently.

In Section 10.3.3, we looked at how we could save interceptors by using SLS instead of salvoing to reduce the interceptor leakage. However, SLS is only an option if the defense timeline allows for a second shot to be taken after the first shot fails. However, if we are launching multiple interceptors anyway to target some decoys that we are not sure of (or because there are multiple RVs in the attack), we can gain many of the benefits of SLS by staggering the interceptors we launch so that later interceptors can observe the results of earlier intercepts and respond to this new information.

In Figure 10.4, we had about 600 seconds of battlespace—the time between the earliest and latest intercepts—and were able to achieve SLS. In Figure 10.18, the endgame time for the seeker is 100 seconds, and it is likely that lesser times could be possible. If we spaced three interceptors by 100 seconds, we would need only 200 seconds of battlespace to have three shot opportunities at different objects. We will call this strategy *ripple fire*. In the simplest case, assume there are two RVs in the threat, and we fire three interceptors separated by 100 seconds. We fire the first interceptor at the first RV and the second interceptor at the second RV. If either of the interceptors misses, we divert the third interceptor to the RV that was missed. Thus with three interceptors, we achieve essentially the same leakage as if we salvoed two interceptors

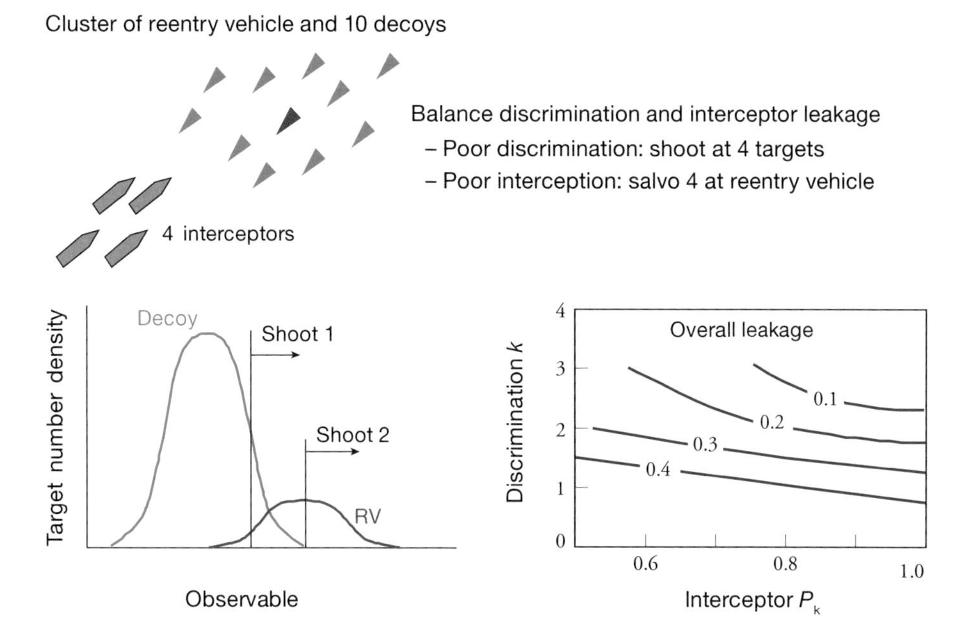

Figure 10.21 What if interceptors are not perfect?

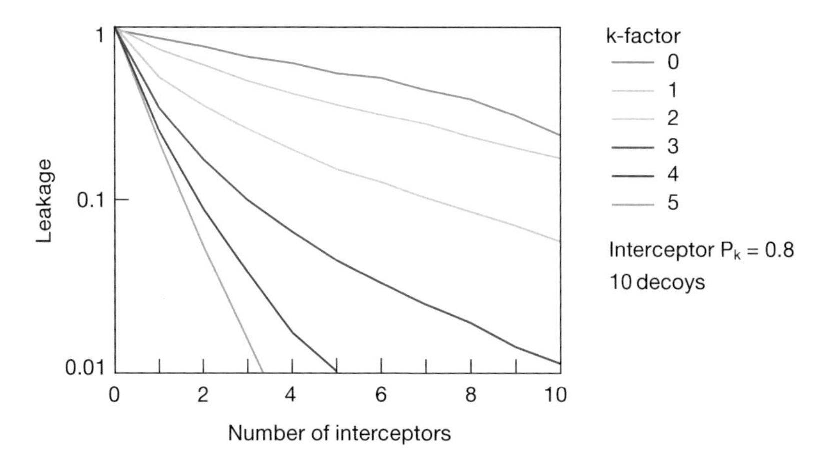

Figure 10.22 Trade interceptors for discrimination and unreliability.

at each of two RVs for a total of four interceptors.

Similarly, if there were one RV but two decoys we were not certain of, we might fire four interceptors in waves. If the first one, targeted at the "sure" RV, misses, we would divert the next interceptor to this RV. If the first interceptor hit the sure RV but it turned out to be a decoy, we would divert the next interceptor to the object next most likely to be an RV. Again, we are achieving a lower leakage with a given number of interceptors. In these cases, as opposed to SLS, we actually fire all the interceptors even though some of them may not be needed. However, we do not need the extended battlespace required for SLS. Figure 10.23 compares the leakage versus the number of interceptors for different firing strategies in different scenarios. The legend indicates the discrimination *k*-factor, the number of decoys, and the interceptor firing doctrine for each curve.

We see that the best strategy depends on the number of decoys and RVs, the *k*-factor, and the kill probability. An effective BMD system should be able to execute all of these firing strategies.

So far, we have only looked at combining the discrimination function with the intercept function. It is also possible to combine the detection function with the other two, but this is a complicated process, and we only discuss it here rather than give

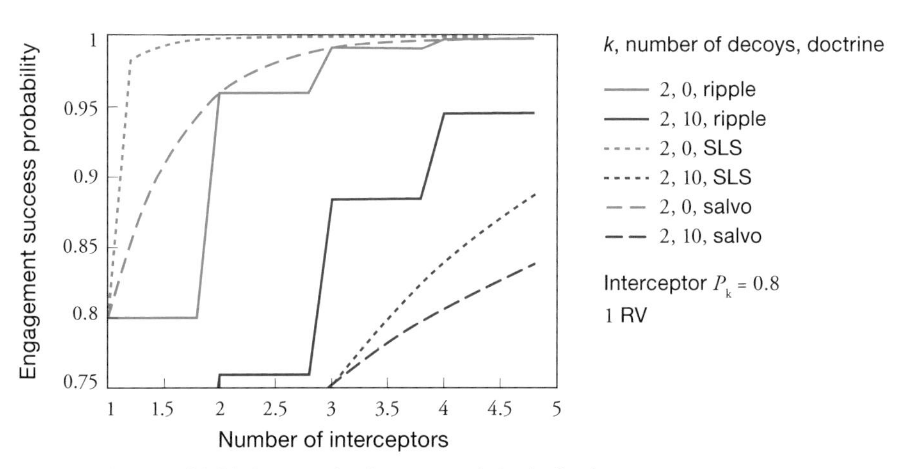

With no decoys, SLS is best, salvo is next, and ripple fire is worst
With decoys, ripple fire is best, SLS is next, and salvo is worst
The ordering will change with number of decoys, RVs, *k*-factor, and P_k

Figure 10.23 Comparison of interceptor firing doctrines.

any results. Detection leakage arises from limited sensor angle and range coverage as well as possible sensor overload if there are a large number of attackers. A sensor must divide its time and resources among the functions of search, track, discrimination, and interceptor handover. The general rule of thumb in balancing these functions is "last things first." If an interceptor is flying toward a target and needs a better handover volume, reducing this volume should be the sensor's first priority. If targets need to be discriminated before it is too late to launch an interceptor, discrimination should be the sensor's next priority. If other targets have been detected but need to be tracked before they are lost, that tracking should be the sensor's next priority. Finally, if all these functions have been met, the sensor should use its remaining time and energy to search for new threatening targets. If a sensor has reduced resources for search, it can cut back in a number of ways, all of which tend to increase the detection leakage. It can search at a shorter range, but this action may result in the non-detection of targets that are attacking the edges of the defended region. The sensor can take longer to search, which may result in some targets getting fewer opportunities to be detected, again increasing leakage. Finally, it may search with a higher detection threshold, resulting in some smaller targets leaking through.

In an operational BMD system, the defense must balance all these forms of leakage and formulate strategies to handle any potential attack. Developing these strategies and evaluating their performance are well beyond the scope of this chapter.

In the final section of this chapter, we discuss some general characteristics of systems analysis that are applicable to BMD systems and many other defense systems. The discussion is a mixture of philosophy and rules of thumb.

10.6 Philosophy and Rules of Thumb

This section presents some rules of thumb about systems analysis in general, with some examples that apply specifically to BMD analysis. It is a mixed bag of topics that hopefully illustrates some important issues. Each topic is discussed separately, but in many cases, the analyst has to apply several or all of them to an analysis.

10.6.1 Systems Analysis Trades off Performance, Cost, and Schedule: Pick Two

For BMD systems as for many other systems, the various options to do the job differ in performance, cost, and schedule. The option selected depends on where in this trade space we have the most desirable (or least undesirable) combination of these variables. The "pick two" part of the section title indicates that we can always satisfy requirements on two of the variables if we completely ignore the third. For example, we can get a system with outstanding performance at very low cost if we are willing to wait 50 years for Moore's law to advance the available electronics technology. We can get a system with outstanding performance right now if we are willing to spend the national debt for it. Finally, we can get a low-cost system right now if we don't care what its performance is. In practice, we must make compromises in this three-dimensional space. Some compromises are considered in the next subsection.

10.6.2 Systems Analysis Is a Fractal[2] Process

There is an old saying that one man's component is another man's system. Generally, the analysis of a system takes place from the top down. We look at the highest-level trade-offs first to select an overall approach to the problem. We then look at each subsystem of the system to conduct further trades and then go through elements, components, and so on. Figure 10.24 indicates the overall process for analysis. Often, system design or synthesis uses a combination of top-down and bottom-up thinking. If we know the state of the art or physical limits for various system elements or components, we can use this knowledge to build up the overall system.

In the plot on the left of Figure 10.24, we can start with the highest-level trade between cost and performance. Here we indicate three options that have different trade-offs in this space. All have asymptotes showing a minimum cost to get any performance at all and a maximum performance no matter how high the cost. Each option has a region of the cost-performance space in which it is better than the other two. When optimizing in two dimensions, we should keep in mind A.J. Liebling's words that "I can write better than anybody who can write faster, and I can write faster than anybody who can write better." Depending on where we want to operate, we could choose one of the options—say, option B. Moving to the center plot, we can expand the performance axis to look at the trade between leakage and coverage. Here we again have three options that differ in their trade-off behavior, and again, the option we choose depends on the relative importance we put on the two performance measures. In this case, we might choose option B2. In the plot to the right, looking at the leakage axis in yet more detail, we might find three options that differ

2 We use the term fractal in the sense that as we look in greater and greater detail, we continue to see more trade-offs in the design and performance of the system. Fractals were originally developed to explain the fact that the length of the coastline of Great Britain increased without limit as the ruler used to measure it grew shorter. Similarly, the analysis of a system can continue without limit unless time or money constraints serve to restrain it.

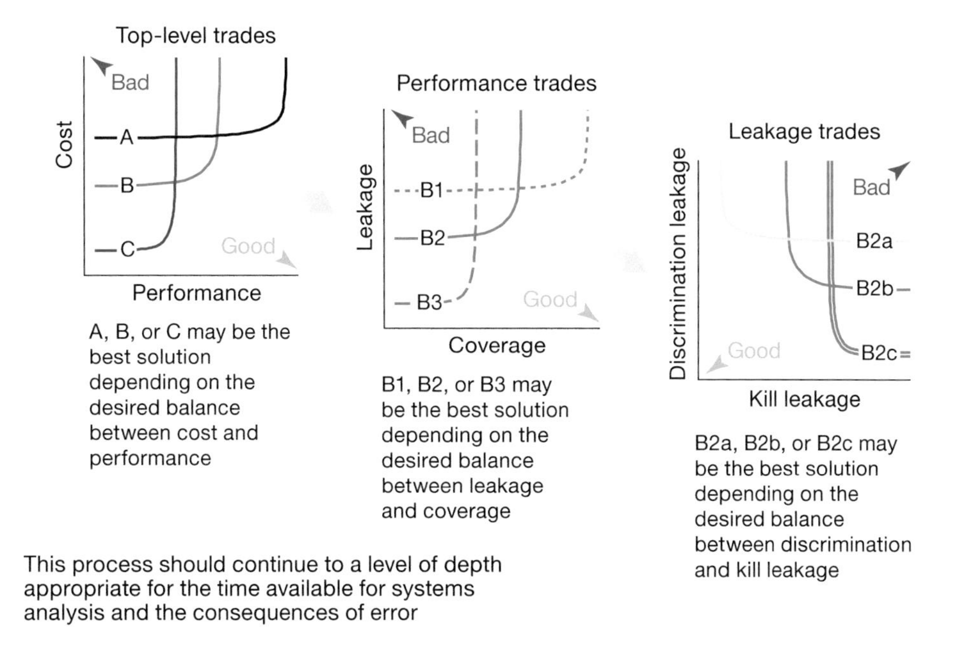

Figure 10.24 The fractal nature of trade-off analysis.

in their trade between discrimination leakage and interceptor leakage. This process can continue through many more levels of analysis. In an ideal world, we would carry all options to all levels of analysis and optimize the design in a 20- or 30-dimensional space rather than selecting a top-level option at the highest level and only continuing this option to lower levels of analysis. However, this comprehensive analysis would take too long and would contain enough uncertainty to call the overall optimization into doubt. Systems analysis contains its own trade-off between speed and fidelity of the analysis. Finding the best path through the tree of trade-offs requires some experience and good judgment. It also requires a little luck. A quotation from the godfather of radar, Sir Robert Watson-Watt, says, "Take the third-best solution; the best never comes and the second-best comes too late." While this advice doesn't always apply in systems analysis, it should be kept in the back of your mind.

10.6.3 Balance Systems to the Extent Possible

In the preceding sections, we have seen numerous trades that allow us to achieve our requirements in a variety of ways. A good example is the sensor-interceptor trade shown in Figure 10.2. We can get a given coverage footprint with a very-long-range sensor and a very slow interceptor, with a very-short-range sensor and a very fast interceptor, or with a medium-range sensor and a medium-speed interceptor. When we put in the cost and risk factors for the very-long-range sensor and the very fast interceptor, we quickly conclude that choosing medium performance elements is better than relying on ultra-high performance for one element to make the other elements easier. This decision is another example of the Goldilocks effect that we saw in Figure 10.17. The specific design point depends on the relative technology levels and costs for different elements, but it is almost always the case that we should try to avoid putting all the risk in one element of the system. As the system evolves, new advances in technology or development risks may change this desired design point, and it is good to retain as much flexibility and performance margin in the system and its elements as possible to permit these system changes when needed.

Another use of the Goldilocks principle is in presenting your results. When giving a spectrum of options, you should generally present three, one of which is too expensive, one of which has inadequate performance, and the third which is just right. Going through this spectrum of options in your analysis is also useful and may give you some ideas for improving the overall system.

10.6.4 Consider Subsystem Interfaces

In Section 10.3.3, we discussed some of the potential problems in handing a target from a radar sensor to an IR seeker on an interceptor. These problems are typical of interface issues that can occur in determining the performance of a system. In the design of the system, interface issues usually concern communication protocols, such as making sure both elements use the same coordinate systems and units and have procedures for dividing up the threat and assigning sensors and interceptors to targets. In analysis, we are looking for problems that might occur if a sensor is not available or is too busy to do a function that other parts of the system assume the sensor is responsible for. We want to see what would happen if the threat is larger or smaller than the system is expecting. Figure 10.25's message applies more to system design than to systems analysis, but it illustrates the type of situation the system analyst should be on the lookout for.

Figure 10.25. An assembly of working components is not guaranteed to produce a working system.

10.6.5 Beware False Assumptions

There are many ways that sensor or interceptor designers achieve greatly enhanced performance by making assumptions about the threat or the environment that may not always be true. Here we give just one example of this: monopulse angle accuracy. A radar has a beamwidth that is approximately the radar wavelength divided by the antenna diameter. The radar can always tell where a target is to within this beamwidth. For example, with a 2-meter antenna at X band (0.03 m wavelength), the beamwidth is about 15 milliradians, or about 1 degree. At a range of 300 km, this beamwidth corresponds to a cross-range accuracy of about 5 km. By forming two receive beams that are slightly offset and comparing the signal strength in these two beams, the radar can locate the target to a small fraction of the beamwidth. Theoretically, the uncertainty can be reduced by a factor of the square root of the signal-to-noise ratio. Often, the signal-to-noise ratio can be 100 or more, giving an accuracy of a tenth of a beamwidth or even less. In our example, the cross-range accuracy using monopulse could very easily be 500 m or even 250 m. However, inherent in the monopulse processing is the assumption that the target is a single point scatterer. If there are multiple targets in the resolution cell, this assumption is not valid, and the results of monopulse processing could be in error by a large fraction of the beamwidth. In this case, we would expect an accuracy of 500 m when the true error would be closer to 5 km. This example is a classic proof of Mark Twain's warning: "It ain't what you don't know that gets you into trouble. It's what you know for sure that just ain't so." Not only does the system designer have to watch out for such questionable assumptions, the systems analyst has to check all the implicit assumptions in the system and find when they are and are not valid.

10.6.6 The Sanity Check

A sanity check is the most important consideration the systems analyst must bring to any problem being studied. Very often, analysis is done using computer models or simulations in which many functions or subroutines are combined and only the final answer is output. This practice is particularly dangerous because a single error in one of the routines may invalidate all the results. Designing an analysis program is, in many ways, analogous to designing the system. Each subroutine should be checked out against hand calculations over a wide range of parameter values to verify that it is doing what was intended. Each subroutine should also be checked to see what it outputs in default cases where some of the input variables represent impossible situations. If one routine outputs 99999 to prevent an infinite output and another routine outputs 0.00001 to prevent a zero output, their combination might output a reasonable number such as 1 when a real answer should be impossible.

In putting together a systems analysis model, we often make simplifying assumptions about the scaling laws that govern the system's performance and cost. As we saw for the radar equation in Section 10.2.2, increasing the radar power by a factor of 16 can increase its range by a factor of 2, all other things being equal. We can get analogous scaling laws for how the cost of a radar varies with some of its design parameters. While these scaling laws may seem reasonable, it is always desirable to compare them with real data on real system elements. Figure 10.26 shows how the combination of data and models can be far superior to either data alone or models alone.

If the data do not agree with your model, you need to understand why not. It may be that the model does not cover the situations that the data describe. It may be that the data points represent a different technology level or are subject to different constraints than the situation for which the model was designed. It may be that the model is wrong. Whatever the reason, it is essential that you understand the reason and be able to explain it simply before presenting your model to an outside audience. This is one important example of doing a sanity check on your analysis.

Other examples of sanity checks are given below. Basically, a sanity check consists of a set of simple calculations that test the consequences of your results. The calculations are simple enough that your audience can do them during any presentation of your analysis. If the sanity check confirms your results, it can lend credibility to the analysis. If the sanity check contradicts your results, it can lead to disaster. Some of the things you can do in a sanity check are to calculate intermediate results. Rather than just calculating the range of a radar, also calculate its power, its beamwidth, its pulse-repetition frequency, its angle accuracy, its signal-to-noise ratio, and any other things that are easy to calculate. Are any of these intermediate calculations silly? For example, do

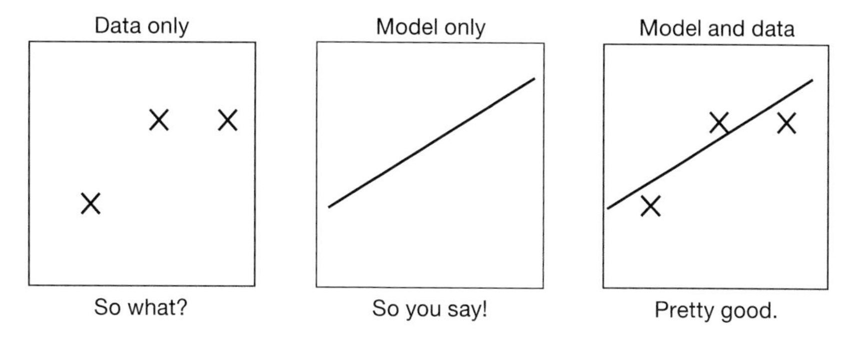

Figure 10.26. Comparison of model with reality.

we have a 1-watt radar with a 100-meter-diameter antenna, or a megawatt radar with a 10-centimeter-diameter antenna? Depending on how we set requirements on search and track, we can come up with such unbalanced designs. Check these intermediate calculations, at least on your final design.

One other form of sanity check is to look at output scaling laws. If you have several options for an interceptor or a sensor, how do their performances and costs scale with their sizes and powers? Generally, a bigger sensor can see further and costs more. Generally, a bigger interceptor can fly faster and costs more. If your analysis shows otherwise, you probably made a mistake. Sometimes this strange behavior is actually correct, given all the constraints on the problem, but again you had better get a good, easily understood explanation for this behavior.

10.6.7 Always Be Willing to Learn Something New

One of the great advantages of systems analysis is that you are always being exposed to new concepts or methods of analysis. Often analysis involves a learning curve as you talk to experts in a field and try to distill their knowledge into those factors that are most important for understanding how this new area influences the system performance or cost or risk. This process will involve sifting through a variety of concepts to find those critical to your analysis, and is half the fun of the work.

Appendices

Appendix 10.A Derivation of the Radar Equation

The discussion here indicates how the range of a radar can be determined from the radar equation, which relates the radar performance to a few radar design parameters. The radar equation is most easily derived by considering what happens to the radar energy for a search radar. If the radar searches a solid angle Ω in a time t with an average power P, at range R, the energy is spread over an area of ΩR^2, giving an energy density of

$$P\,t\,/\,\Omega\,R^2.$$

The radar cross section of the target, σ, is defined as the equivalent area of a sphere that would reflect the same energy back to the radar as the target does. Since a sphere scatters energy equally in all directions, the energy density back at the radar is

$$P\,t\,\sigma/\,4\,\pi\,\Omega\,R^4.$$

The received energy is this energy density times the antenna area A. This represents the signal. The signal is reduced by the losses in the system, L, and compared with the noise (usually stated as Boltzmann's constant, k, times the noise temperature T.) Thus, we have a signal-to-noise ratio

$$S/N = P\,A\,t\,\sigma\,/\,4\,\pi\,\Omega\,R^4\,kT\,L. \tag{10A.1}$$

Boltzmann's constant is 1.38×10^{-23} joules per degree, and the fact that it is so small is what makes long-range radar possible. Equation (10A.1) can be rearranged to put the performance parameters on the right-hand side and the design parameters on the left-hand side:

$$P\,A\,/\,kT\,L = 4\,\pi\,\Omega\,R^4\,(S/N)\,/\,\sigma\,t. \tag{10A.2}$$

Making the right-hand side bigger makes the radar more capable; making the left-hand side bigger makes the radar more expensive. As with interceptors, we have the option of deploying more lower-performance radars or fewer higher-performance radars. Another quantity-quality trade-off.

Equation (10A.1) can also be used to determine the tracking performance of a radar. Rather than searching a solid angle, a tracking radar needs only to send energy where the targets are. The best that an antenna can focus the radar beam is a beamwidth of ~λ/D, where λ is the radar wavelength and D is the antenna diameter. The time that the radar can spend on each target is one over the measurement rate or the pulse-repetition frequency (prf). To get from Equation (10A.1) to the radar track equation, we replace Ω by the square of the beamwidth and replace t by 1/prf. The resulting track equations are

$$S/N = P\,A^2\,\sigma\,/\,4\,\pi\,\lambda^2\,\mathrm{prf}\,k\,T\,L, \qquad (10A.3)$$

or separating performance and design parameters,

$$P\,A^2\,/\,\lambda^2\,k\,T\,L = 4\,\pi\,R^4\,\mathrm{prf}\,(S/N)/\,\sigma. \qquad (10A.4)$$

We can also use the definition of antenna gain, $G = 4\,\pi\,A/\lambda^2$, to rewrite Equation (10.4A) as

$$P\,A\,G\,/\,k\,T\,L = (4\,\pi)^2\,R^4\,\mathrm{prf}\,(S/N)/\,\sigma. \qquad (10A.5)$$

The power-aperture-gain (PAG) product is often used as a measure of track radar capability. It is certainly a measure of radar cost.

When I first started working on BMD, Bill Delaney told me that if I understood the radar equation, I would be able to pay off my mortgage, and if I further knew how to spell radar backwards, I could put my kids through college. He was right.

Appendix 10.B Derivation of the Optics Equation

Here we derive some scaling laws for the range of an optical or IR sensor. These sensors are passive and respond to radiation emitted by the target as well as to radiation emitted by background sources both internal and external to the sensor. For confident detection and track, the signal from the target needs to significantly exceed the noise from the background. Both signal and background depend on the temperature of the objects giving off the IR radiation. For detection of room-temperature targets, the sensors need to have cooled optics and focal planes and to view the targets against a space background. Here we do not develop exact equations for the IR signal-to-noise ratio, but will indicate how the ratio scales with parameters such as search solid angle, search time, range, sensor aperture, focal plane size, and background and signal levels.

To derive an optics equation, we use the resolution or field-of-view variables in angular space that were indicated in Figure 10.7.

The signal collected by the sensor is proportional to the target intensity in the sensor wave band I_t, the sensor aperture area A, and the integration time τ, and inversely proportional to the square of the target range R. The mean background B is proportional to the background intensity per unit solid angle I_b, the aperture area, the square of the pixel instantaneous field of view (IFOV), and the integration time. Usually, the mean background level is significantly higher than the signal level. However, these sensors detect targets by looking for pixels that are significantly higher than their neighboring pixels. In this case, we are not concerned with the mean background level but rather with the variations in the background. For the pixel containing the target, the level will be the mean background, plus or minus the fluctuations in the background, plus the signal level. For the neighboring pixels, the level will be the mean background plus or minus the fluctuations in the background. For most types of background, the fluctuations in the background are proportional to the square root of the mean background. With these assumptions, we can determine an effective signal-to-noise ratio scaling for the pixel with the target having the following dependence on target and sensor parameters:

$$S \sim I_t A \tau / LR^2 \tag{10B.1}$$

$$N \sim \sqrt{B} \sim \text{IFOV} \sqrt{I_b A \tau / L} \tag{10B.2}$$

$$S/N \sim I_t \sqrt{A\tau} / \text{IFOV}\, R^2 \sqrt{L\, I_b} \tag{10B.3}$$

Combining this with the expression for τ from Figure 10.7, we get

$$S/N \sim \mathrm{I}_t \sqrt{Q\,A}\,\sqrt{t\,/\,\Omega}\,/\,\mathrm{R}^2 \sqrt{L\,I_b}\,. \tag{10B.4}$$

In these equations, we have assumed that the losses L are the same for the signal and the background. In some scenarios, they will have different loss factors. Equation (10B.4) has similarities and differences with the radar equation, (10A.3). These can be seen more clearly by rewriting Equation (10B.4) with design parameters on the left and performance parameters on the right:

$$Q\,A\,/\,L\,I_b \sim (S/N)^2\,\Omega\,R^4\,/\,t\,I_t^2\,. \tag{10B.5}$$

In this form of the equation, on the left-hand (design) side, $Q\,A$ plays the role of the power-aperture product and I_b plays the role of the noise. Again, increasing the left-hand side makes the sensor more expensive. On the right-hand (performance) side, the dependence on Ω, t, and R is the same as in the radar equation. However, the dependence on the target intensity I_t and S/N is quadratic rather than linear as in the radar equation. This is not surprising since the sensor types are very different. More surprising are the similarities.

About the Author

Stephen D. Weiner is a senior staff member in the Systems and Architectures Group at MIT Lincoln Laboratory. He received SB and PhD degrees in physics from MIT in 1961 and 1965. Since then, he has worked at Lincoln Laboratory in many areas of research in ballistic missile defense as a staff member, group leader, and senior staff member. He has worked in areas of system and sensor design, sensor tracking and discrimination measurements, and interceptor guidance. He has also written several book chapters on the history and technology of ballistic missile defense. He has participated in a number of national studies looking at defense against both theater and strategic ballistic and cruise missiles. Dr. Weiner is the recipient of a 2004 Lincoln Laboratory Technical Excellence Award. In 2008, he developed the concept of a negative Technology Readiness Level.

11

Systems Analysis in Nontraditional Areas

Michael P. Shatz

11.1 Introduction

I came to Lincoln Laboratory in 1984 after finishing a PhD degree in particle physics at Caltech. I joined a group doing systems analysis for the Air Force Red Team, as Aryeh Feder describes in Chapter 6. My doctoral dissertation was a detailed calculation relating to experimental tests of quantum chromodynamics, but my finely honed skills at complex and difficult calculations were not the facet of my education that helped me most in my career as a systems analyst at the Laboratory. In fact, except for a couple of forced attempts in my first year or two, I never had to do mathematically challenging calculations to do well-regarded work. Instead, what mattered most were skills I had learned without realizing I had learned them, mostly from attending a wide range of seminars and colloquia and seeing how the best speakers explained their work and how the faculty interacted with the speakers. Those skills were an ability to quickly assimilate new technical material and step back from it to see the main ideas, and the ability to set up mathematical models to quickly assess the implications of the ideas. These skills were refined by doing systems analysis under the tutelage of experienced hands like Alan Bernard and John Fielding.

I was fortunate to be at Caltech when rapid advances in instrumentation and computation were making possible dramatic progress in multiple areas of physics, and when physics was overlapping with other sciences like biology. I was particularly fortunate to have the opportunity to be a teaching assistant for Nobel Laureate physicist Richard Feynman. He was a master at both making detailed calculations and stepping back to see the big picture. The importance of stepping back was brought home to me very forcefully on one occasion in my graduate career when I consulted Prof. Feynman on a difficulty in the calculation I was then pursuing for my thesis work. Prof. Feynman helped with the immediate difficulty, but then promptly suggested that I find another thesis topic. He offered two reasons. The first was a technical issue he thought would arise in the next level of the calculation. The second and more important reason was a simple argument by dimensional analysis that even if I succeeded in getting the calculation to work, the effect would be impossible to discern in experimental data from other equally large effects at the energy scale of relevance—indeed, a whole infinite series of them. I spent two or three weeks working very long hours trying to do the next-order calculation and trying to work through (well, to be honest, refute) the dimensional analysis argument before convincing myself that at least the dimensional argument was completely right and finding another thesis project. Since that painful lesson, I have always tried to do at least a quick estimate of what I expect the answer to be before embarking on a lengthy, detailed calculation. This approach is something I have recommended to analysts whenever the occasion arose, and I have never regretted it.

11.2 Nontraditional Areas

This chapter addresses the use of systems analysis in new mission areas. After 10 years or so doing air defense analysis for the Air Force Red Team, I led the Lincoln Laboratory portion of an effort to look at countermeasures to precision-guided munitions. After the dramatic successes of such munitions in the First Gulf War made the world aware of their capabilities, the Air Force became concerned that potential adversaries would attempt to develop such countermeasures. After a year or two of this work, I was asked to help out with a study of surface surveillance missions for an effort sponsored by the Defense Advanced Research Projects Agency (DARPA). Surface surveillance was an area in which the Laboratory had technological expertise but no systems analysis capability and little understanding of the missions; our sponsors were mostly technology offices that also had only a vague understanding of the missions. This study, described in more detail in the next section, led to the formation of a new group chartered to do systems analysis in new areas. The group's portfolio eventually included homeland protection and defense against bioterrorism. This chapter focuses on how to get information and quickly perform useful analysis in a new field; it also includes some of my experiences dealing with audiences that were not as well primed as the Air Force Red Team to value systems analysis.

11.3 Surface Surveillance

Around 1997, I was asked to look at a study to support an effort at DARPA to develop a large constellation of space-based radars that would survey Earth's surface. The idea was to have persistent, near-constant access to some sort of surveillance data anywhere in the world. To make the system affordable (or some approximation thereof), the satellites would only look at a small fraction of their potential field of regard in each pass, and produce either low-resolution synthetic aperture radar (SAR) images or ground moving target indicator (GMTI) hits, perhaps with relatively coarse range resolution. The study was to quantify the utility of such data to military missions in order to determine how design trades could maximize the utility of the system for a given cost and to justify that cost in terms of the benefits received. Over the ensuing years, this study would prove to be one of several programs that looked at the tactical value of what has come to be called persistent surveillance. Although these programs did not directly result in deployments of such systems, the ideas generated did bear fruit in a variety of forms several years later.

Evaluating the military utility of this type of surveillance led to a type of systems analysis very different from my previous work, which had been on air defense and on countermeasures to precision-guided munitions. In those analyses, the basic issue was the detection of a signal in the presence of interference. Such detection was based firmly in the physics of the strength of the signal and interference, and in statistical detection theory to compute probabilities of detection and false alarm. The challenge of the analysis was to use these well-defined theories to gain insight into the operational problems at hand. The question of the utility of ground surveillance data at different resolutions and frequencies focuses on the ability to identify and track targets of interest and separate them from objects of lesser interest (clutter or confusers). This problem space is one in which there was not (and still is not) a complete theory on which to base analysis; truth be told, there wasn't even much in the way of a useful incomplete theory. However, the systems analyst was not left to operate blindly. In that period, a variety of experimental data shed light on some of the key issues; also, well-documented procedures allowed military analysts to use the ground surveillance they were accustomed to getting. These data and procedures formed a basis for analysis that illuminated the potential benefits of persistent surveillance and the design trades for the required systems.

Classifying types of targets for GMTI data was an early issue some analysis was able to clarify. One thing that provided important insight was looking at the requirements to track military units of various sizes. We were able to show that it was much easier to track large units than small ones, and to quantify by how much. Of course, as in most good analysis, the result is obvious in hindsight: larger units cover more ground, are harder to confuse with other things, and move and maneuver more slowly than smaller ones. Table 11.1 was presented in an early talk showing revisit times needed to track different-sized units. These results could be put into a mission context by using the then-current Army field manuals. The field manuals described what types of action the opposing forces were likely to take, what areas and sequence they might be in, and what level of command (e.g., theater, corps, division) might want to track them. This information could be used to get an estimate for what kind of data rate would be needed in different theaters to support a given level of command.

In addition to looking at military units, we also looked at individual vehicles. Early in the effort, the sort of vehicle we had in mind was a missile launcher; this decision was motivated by the Scud-finding problem from the first Gulf War. Missile launchers are fairly distinctive even if all one has is a length measurement or a one-dimensional reflectivity profile. We also gave some attention to less distinctive vehicles; the concern at the time was a situation such as a panel truck leaving a factory suspected of producing weapons of mass destruction. This vehicle class became more important as terrorism came to the fore after 9/11, and there was interest in tracking particular passenger vehicles. There is often a trade between the frequency of observation and the resolution with which a sensor can measure targets in an observation, that is, between

Table 11.1 Required revisit rate for tracking military units by using perceived enemy doctrine.

Force	Number of fighting vehicles	Travel rate	Area covered by unit (km^2)	General maneuver execution time	Revisit time needed
Army	>3000	Open even terrain: 100 km/day Paved roads: 40 km/day Uneven terrain: 20 km/day	5000–10,000	4 hr	30 min
Division	~1000	Open even terrain: 150–200 km/day Paved roads: 50–60 km/day Uneven terrain: 40 km/day	500–1000	1.5 hr	10 min
Regiments	~350	Paved roads: 100 km/day	50–100	30 min	3 min
Battalions	~120	Paved roads: 80 km/hr max	5–10	10 min	1 min
Company	~40	Paved roads: 80 km/hr max	1–3	5 min	30 s

Logistic vehicles not accounted for. Rule of thumb: ~2× number of fighting vehicles.

the frame rate and the ability of the sensor to discriminate the desired targets from other vehicles. The two extremes in this trade are easy to analyze but, at the time, such sensors were impossible to build:

- A sensor with the resolution to read the vehicle identification number through the windshield would be able to operate with a low frame rate but would never lose track of the vehicle as long as it could keep its field of view around the target.
- Likewise, a sensor that could detect the target with video frame rates would not need to discriminate the target from other vehicles; because vehicles don't move their lengths in one frame, the sensor only has to be able to tell a single vehicle from two, one passing the other.

At that time, however, real sensors had to operate in the uncomfortable region between the two extremes, with resolution that supported only imperfect target recognition and revisit rates too low to keep track unambiguously by the kinematics.

Back then, sources of data for accurate models of traffic in the operational contexts of interest were not really available, so we began with very simple models. We tended to assume simple kinematics, such as vehicles with speeds as Gaussian random variables about a nominal speed placed at random along a linear road, or vehicles that moved randomly like a two-dimensional ideal gas. An example of a simple model was to assume a given density ρ of vehicles moving at random. If A is the area the target could move to between looks, there will be on average $A \bullet \rho$ targets to confuse with the one we want. Thus if f is the fraction of false targets that can be removed per look by some combination of better assumptions about the dynamics of the target and automatic target recognition (ATR), then depending on whether f is greater or less than $A \bullet \rho \,/\, (1 + A \bullet \rho)$, the number of targets that could be the one desired either approaches a constant or grows exponentially with the number of steps. You might think that this analysis is too crude to be of use, but it did offer some important insight. One contributing factor is that unless f is very close to the boundary value, the number of possible targets either grows very fast or approaches a constant close to 1. Therefore, without being terribly sensitive to the details, it is possible to tell whether a given combination of false-target rejection rate and frame rate is likely to result in an easy tracking problem or one that is nearly impossible. We were later able to get some more sophisticated traffic models to compare these calculations against, for example, a model of traffic in American cities used by traffic planners. Using these more sophisticated models tended to change the detailed numbers but did not change which problems were easy, which were on the edge of possible, and which were just too hard for the contemplated technology.

At that time, we tended to assume that the only ATR algorithm available to a GMTI radar was to measure projected length to an accuracy set by the range resolution of the radar. We did have estimates of the numbers of vehicles of various lengths in different countries; this information allowed us to estimate the performance of the ATR algorithm rejecting a confuser vehicle. We also had a fair amount of data on the ability of SAR imagery to distinguish among various types of military vehicles. Analysis based on this limited data does not, of course, produce numerically accurate results for system performance. But it did provide useful insights into which problems could be addressed by MTI radars and into the trades between radars tuned for MTI modes and those tuned for SAR imaging. The key point is to compare the rate at which ATR can dispose of potentially confusing tracks with the rate at which kinematics and traffic density present possible confusers. If the ATR can dispose of more confusers per measurement than kinematic confusion adds, it is possible to maintain a track. The analysis showed that for the frame rates contemplated at the time (10 seconds and up), one needed ATR performance from 70% to more than 99% for plausible vehicle densities. The ATR performance needed to track single vehicles with MTI radars of the types being considered was possible only for vehicles that had unique signatures and did not operate in dense regions. This analysis helped focus GMTI radar proposals on formations and on very distinctive vehicles like missile launchers. After the initial calculations, more involved calculations were done using more sophisticated models of traffic motions (some coming from studies of traffic used for road planning), models of specific sensors, and ATR algorithms, but the trends illuminated by simple models held up well. This outcome is typical; the role of good systems analysis is to provide insight into the big drivers of a problem, not to calculate highly accurate details. While the first-order analysis necessarily gave approximate numbers, both for the level of kinematic confusion given a data rate and for the ability of ATR to help reject false tracks, it did give reasonable estimates.

The analysis made intuitive sense; in fact, I confirmed the basic insight by a little experiment when I was staying in a suitably high hotel room in the Washington, D.C., area. I looked out the window and opened my eyes briefly every 15 seconds or so. It was impossible to track a specific vehicle of very common types, such as taxis and passenger sedans, but was trivial to track a crane or tanker truck as long as it remained in view. This type of quick experiment, or a quick analysis of historical data, can often be a valuable sanity check. More accurate models of kinematics and ATR preserve that basic insight while providing better estimates of the boundary between the cases and the system performance for cases near the boundary.

About a decade later, this surface surveillance analysis became of reduced importance as rapid advances in visible cameras generated by the cell phone industry, coupled with advances in processing and communications, eventually led to another approach to persistent tracking of individual vehicles using very-large-format cameras on aircraft, at least for cases in which obscuration by clouds is not a problem. This approach, not yet possible when this work began, does come close to providing video frame rates while at the same time providing reasonably high-resolution data, at least over modest (city-sized) areas in theaters where air defenses are not an issue. We expect analysis of radar surveillance systems of the type considered in the above analysis to become important as the United States again begins to consider peer conflicts in which the greater standoff afforded by radar will again be important.

Another area that generated attention in the period was the issue of radar beams and modes. The types of radars under consideration at the time were combinations of GMTI and SAR systems. The question of how much electronic beam steering to include was also a major issue. How to balance the modes and use radar resources effectively were contentious issues at the time. GMTI from the Joint Surveillance and Target Attack Radar System (JSTARS) had been used effectively in the first Gulf War, mostly by tactical operations soldiers, while SAR images had been the province of intelligence users. Although there had not been much thinking about combining and balancing these different types of data, the analysis we will discuss now helped spark and illuminate some of that thinking. Typically, the analysis would proceed by looking at several strategies for accomplishing a given mission and comparing the load on the system. An early example is shown in Figure 11.1, which compares performance in a

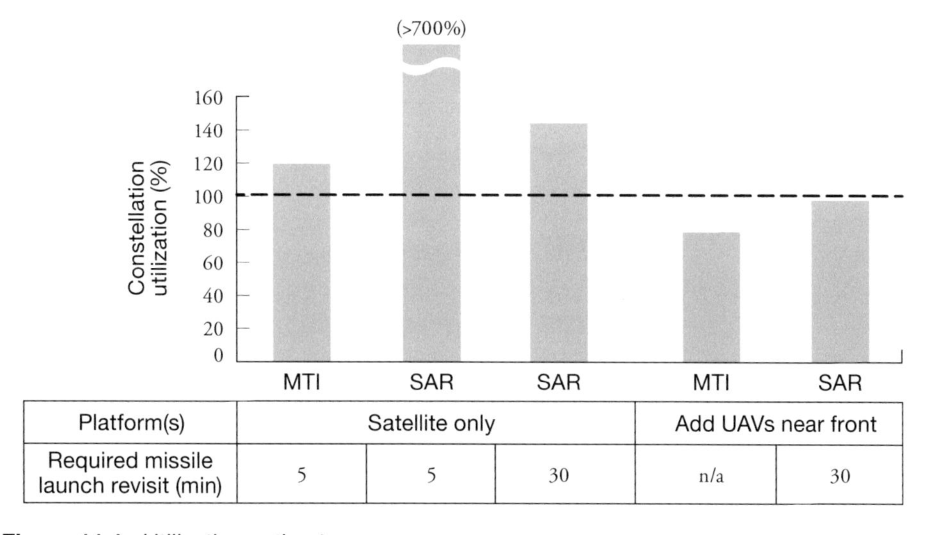

Figure 11.1 Utilization estimates.

scenario that requires finding missile launchers and tracking military formations. To calculate that result, we examined needs for information on enemy formations and missile launchers, and developed strategies to collect as much of that information as possible. The figure compares a strategy primarily using MTI to track all targets, using SAR only to image suspected missile launchers that have stopped, to one that searches SAR imagery to find stationary launchers. Because the latter so overstretched the system, we also looked at a strategy that provided less information than desired by reducing the frequency of SAR searches. Finally, we showed what could be done by using radars mounted on unmanned aerial vehicles (UAV) to search near the front. We showed that the full set of tasks overwhelms the satellite constellation, but that enough of the high-revisit tasks are close enough to the border to allow airborne surveillance to effectively supplement the satellites.

The analysis required for issues of utilization and beam diversity was of a different character than many other types of systems analysis. Because there is no theory allowing one to calculate the amount of information that can be derived from an image or range-Doppler profile, the best we were able to do was develop several possible strategies to accomplish given military objectives with data at various resolutions and estimate the level of radar resources (and, sometimes, analysis resources) needed to implement the strategies. The radar resources used could then be compared with the utility of the resulting information. Absent a theory, there is no way to tell how close any of these strategies might be to optimal, and it is always possible a cleverer person could devise a more efficient strategy. Nonetheless, examining multiple strategies did allow us to gain insights into the relative utility of imaging and MTI modes for particular problems, a method to efficiently combine them, and a way to make efficient use of systems that offered particular combinations of modes. Likewise, the beam steering results depended on assumptions about the geographic distribution of both targets and sensors; however, we were able to derive sensible insights into the degree of beam agility that was valuable in most scenarios, and the results seemed to be reasonably insensitive to the particular assumptions.

It took a while for the types of analysis described to begin to have an influence beyond the immediate sponsorship of the effort, but they did influence a number of radar surveillance and data exploitation programs for DARPA and the Air Force and, to a lesser extent, the Army. As attention shifted to counterinsurgency and as the technology of UAVs and small, high-pixel-count cameras developed, surveillance using close-in video cameras became more important than the radar systems we had been investigating. This was a case of failing to take my own advice about stepping back from the problem, and I was slow to realize that the technology had changed significantly and react accordingly.

11.4 Defense against Bioterrorism

In the mid-1990s, the DoD increased attention to defense against biological attacks. In response, Lincoln Laboratory developed some important technologies for environmental monitoring and sample collection. The anthrax attacks in the United States in October 2001, in which letters containing anthrax spores were mailed to select recipients, motivated increased attention to biological terrorism in urban settings. In early 2002, the White House office that preceded the Department of Homeland Security (DHS) asked the Laboratory to investigate architectures for defense against urban bioterrorism. This investigation required significant systems analysis both to identify the key problems to be addressed and to develop the architectures needed to address them.

Our systems analysis in this area was made more difficult by several factors. In the first place, we had on staff systems analysts whose education and expertise were in physics and engineering rather than biology, and biologists who had little understanding of systems analysis. Second, the potential range of threats is enormous, with numerous choices for agents, both natural and genetically engineered, delivery methods, and scales of attack. Third, the response would be in the hands of a public health system with a culture geared toward responding to natural disease outbreaks not to large-scale terrorist attacks. Finally, there are a great many potential attacks for which no known solution is possible. We analyzed these problems in a broad initial survey and followed up with a variety of more focused analyses over the succeeding years.

When we began this investigation, a variety of studies had used different criteria to examine and prioritize threat agents. One study had been done at Lincoln Laboratory, another by the Centers for Disease Control and Prevention (CDC), and others by intelligence agencies. These studies came to different conclusions, but there was enough common ground to enable a systems analyst without much knowledge of medicine and biology to get started. In the first place, the studies all made clear that there were a great many potential threats, so a system too tightly tuned to any specific threat would prove very vulnerable. Second, anthrax is a very good choice for many kinds of large-scale outdoor attacks because of its ease of acquisition and its environmental stability. Third, there are important agents for which no good medical treatment exists, so it is valuable to be able to respond wherever possible to limit or prevent exposure. Fourth, there are far too many plausible agents for a so-called "one bug one drug" approach to work. The lists of potential threat agents generated by the studies also illustrated an important difference between the approach of a systems analyst and that of a technology developer. Technology developers tended to view these lists as establishing priorities for developing immunoassays, polymerase chain reaction (PCR) primers, vaccines, or medicines; for these purposes, the differences among the competing lists were a problem. For the systems analyst, the variation among the lists was a measure of how much uncertainty the systems had to be designed to accommodate; the common features gave us an initial place to focus priorities.

One early group of analyses, therefore, looked at ways of avoiding exposure. We looked at both airborne attacks and waterborne cases. For the airborne cases, we looked at both residential buildings and at more modern office buildings, which typically have centrally controlled heating, ventilation, and air conditioning (HVAC) systems. For the residential buildings, we looked at closing windows and doors before the threat cloud gets there, and opening them after it passes. What we found was that most residential buildings are leaky enough for the door/window closing to have little benefit; furthermore, if closing and reopening the doors and windows aren't timed right, the strategy has the potential to do harm as it is possible to keep contaminated air in the building longer than needed if the windows are not shut before the cloud gets there and are not opened at the right time. Figure 11.2 shows that keeping the windows shut when the cloud has passed can retain aerosol in the building's air. What happens in this situation is that, even though the concentration of the aerosol is initially reduced by closing the windows, the aerosol that leaks in can stay in the building for longer, resulting in a larger total dose.

On the other hand, in buildings with modern HVAC systems, it is generally possible to keep out contaminated air by overpressuring the building with filtered air and increasing the rate at which air is filtered to remove aerosol particles. This technique could be implemented with local sensors or remotely controlled, centrally operated sensors. Furthermore, there is some tolerance for false alarms, as the HVAC

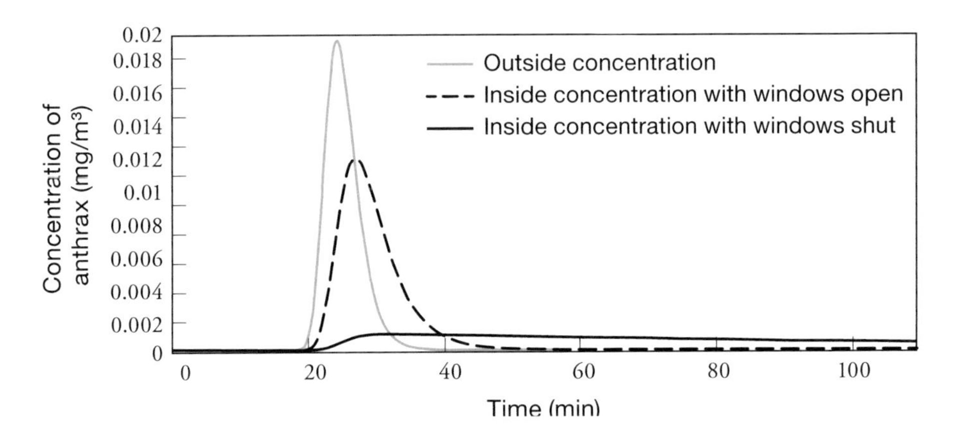

Figure 11.2 Concentration curves for anthrax under three conditions.

changes would not be readily apparent to the building occupants, thus limiting the possibility of panic. A test bed to test this class of systems was implemented at the Lincoln Laboratory facility, and limited operational systems were deployed.

To look at water-based attacks, we focused on Boston's Massachusetts Water Resource Authority (MWRA) system, for which the MWRA provided a hydraulic model. This model enabled us to determine where the system was vulnerable to attacks and how much time there was from the time of the injection of agent into the system until it became impossible to shut off the water or switch to water from an uncontaminated backup source before it reached consumers. We were thus able to define timelines required of sensors to prevent possible exposure. The cost estimates we got for many responses to alarms enabled us to define break-even false-alarm rates (that is, the false-alarm rate at which the cost of responding to false alarms is equal to the expected benefit from earlier response to true attacks). While we did have to learn something about the survival of organisms in a treated water system, the detailed knowledge we needed most was not biological information but the engineering detail of the water system.

This theme comes up often in systems analysis in new areas; when we start to do some analysis, we often find that the detailed domain knowledge we need is from a field different from what we had initially thought. The waterborne attack analysis was an example of how systems analysis can illuminate a new problem and define useful direction for technology programs; we were able to define a cost-effective way to mitigate large-scale attacks. Indeed, we did briefly work on a technology development follow-on program for sensors to meet the demands we had defined. However, it also proved to be an example of how working on a new problem in an area where policy is rapidly evolving can be frustrating. Shortly into the follow-on program, federal responsibility for protecting the water supply was transferred from the then-nascent DHS, which had sponsored the work, to the Environmental Protection Agency, with which we were unable to gain much traction.

However much one might try to avoid contamination, many attacks could unavoidably expose many people to biological agents; for example, a large outdoor aerosol attack will almost certainly expose large numbers of people who are caught outdoors when the aerosol is released or when the cloud passes by. In those cases, sensing can detect the attack early and speed effective treatment to those exposed. This scenario was addressed in a series of studies for the DHS. The architecture then in place was highly dependent on medical diagnosis to start any substantive action; the BioWatch program of environmental sensors was being deployed, but the public health community was unsure how to respond to a confirmed positive. The analysis began by looking at the steps needed for mass distribution of antibiotics. The main steps involved bringing the needed supplies to the area, setting up distribution centers, manning them, and getting the public to show up, wait their turn, and use the medicine appropriately. Each of these steps takes time, and the effect of delay can be severe because the effectiveness of the medicine is usually greatest early in the course of the disease; in the case of anthrax, it is most effective before the patient exhibits symptoms.

Our first analysis looked at simply how many people could be treated with the antibiotics before becoming symptomatic, given assumptions about the disease progression and the rate at which the various steps could happen under different scenarios. Of course, later modeling looked at timelines more carefully and accounted for better models of the effects of treatment at different times, but the majority of the insight came from the original simple model. I have found that it is often the case that most of the insight comes from simple models, and more subtle models are useful for quantifying either performance or requirements more accurately.

The question of disease progression raises another concern that was common in the analysis of biodefense problems. Little is known about the progression of inhalation anthrax, which was one of the main agents we studied, in humans. (The same is true for most other agents we studied.) We had two main data sources to shed light on this problem. One was the results of an accident in a Soviet weapons plant in Sverdlovsk in 1979, and the other was the October 2001 mailings in the United States. The two events showed different profiles for when the exposed became sick, but, of course, neither event was well instrumented. However, the data seen in Figure 11.3 show that in the United States, those who became sick developed their symptoms between four and six days after exposure; in Sverdlovsk, the first illnesses were two days after the release, and new cases continued for 40 days. It is entirely unclear whether these differences are because of different strains, because of dosage-dependent responses, because of some later cases in Sverdlovsk coming from exposure to re-aerosolized agent rather than the initial release, or because of other factors. There being no ethical way to resolve these uncertainties experimentally, the analysis must live with this level of uncertainty. The different incubation models show the importance of distributing antibiotics rapidly; however, they do put different values on systems that can't meet the very aggressive requirements to get the whole population on antibiotics before the first symptomatic cases. While there were (and are) plans for the various phases of the deployment, they have thankfully never been put to the test so there is considerable uncertainty about how well they would work in practice, especially given the strong possibility of panic among both health-care providers and the public.

The initial study examined different ways of cuing the transport of antibiotics, setting up the distribution, and initiating the distribution. The baseline case is that everything is done after medical diagnosis, which can occur only after the first patients

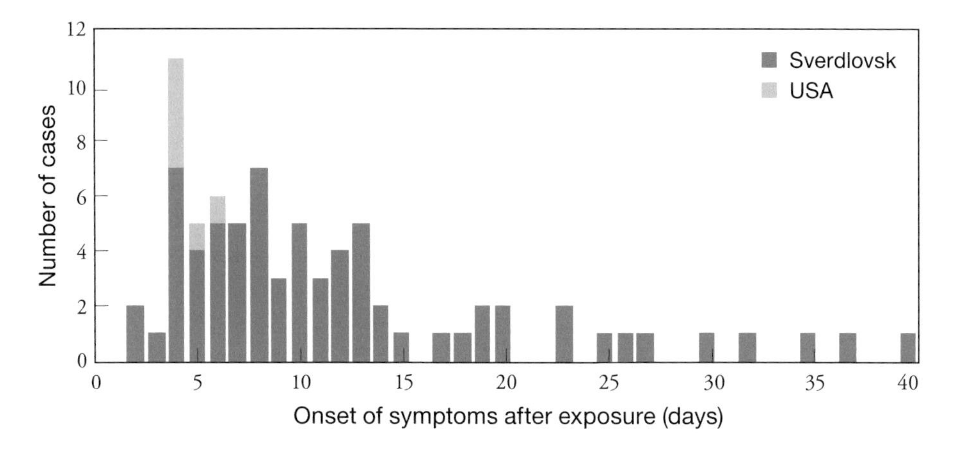

Figure 11.3 Histogram of the incubation periods for anthrax based on data sets from two outbreaks.

become symptomatic, and which probably occurs more than a day later. Because the initial symptoms of almost any disease contracted by inhalation are nonspecific signs of inflammation in the respiratory tract, doctors do not usually look at once for the agents likely in a terrorist attack. The first thing that can be done in response to an environmental alert is to notify the medical community to look for signs of the suspected agent in their patients; in consultation with medical experts, we estimated that this alert may speed up the process by about a day. We also considered transporting supplies from the strategic stockpile in response to either an environmental alert or a confirmed environmental detection. We can go further and also set up distribution centers or begin distributing antibiotics in response to environmental detection. All these steps would speed distribution, and thus reduce casualties, but at the price of increasing the steps that would be taken in response to an environmental false positive.

The benefits of rapid response were analyzed in a series of studies. Figure 11.4 shows the estimated deaths from a large attack, for which we would expect 50,000 fatalities if no one were treated, as a function of how the stages of the response (transporting the supplies, setting up distribution centers, and distributing the antibiotics) are started. The graph shows that the current practice of responding to uncued medical diagnoses (i.e., the physician identifying the agent without any alert to look for such an agent) would limit the deaths to about 17,000 and that if environmental cues are used more aggressively to start more stages of the response, the fatalities decrease but remain very high. These cues can be used for anything from alerting emergency-room physicians

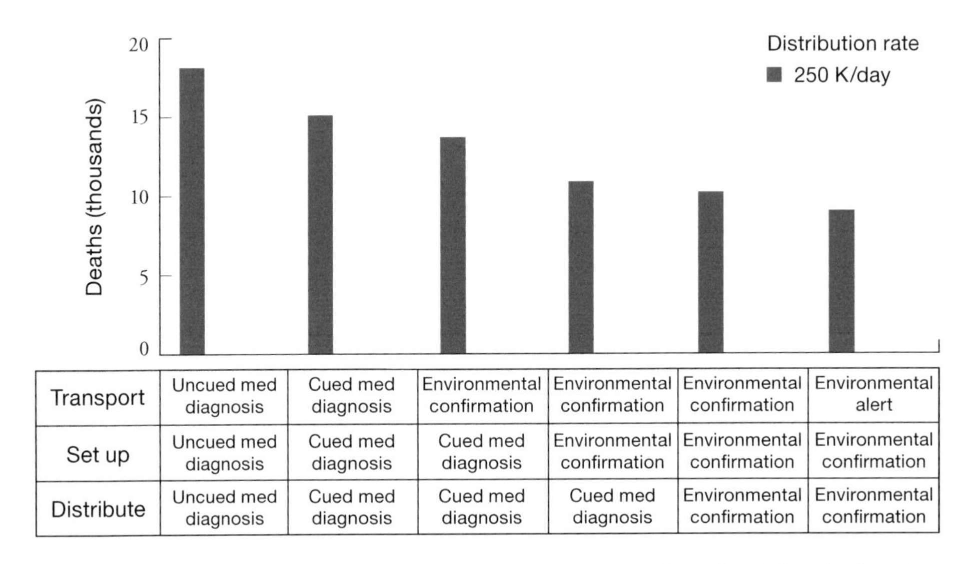

Transport	Uncued med diagnosis	Cued med diagnosis	Environmental confirmation	Environmental confirmation	Environmental confirmation	Environmental alert
Set up	Uncued med diagnosis	Cued med diagnosis	Cued med diagnosis	Environmental confirmation	Environmental confirmation	Environmental confirmation
Distribute	Uncued med diagnosis	Cued med diagnosis	Cued med diagnosis	Cued med diagnosis	Environmental confirmation	Environmental confirmation

Figure 11.4 Performance of different architectures given the Sverdlovsk incubation data.

to look for the agent used in the attack in patients with nonspecific symptoms, to positioning resources drawn from the National Strategic Stockpile, to beginning mass distribution of medicines before a diagnosed case. Because of the time required to distribute antibiotics to the entire population, thousands of people who were exposed would not receive them in time for the treatment to be effective. The calculation assumes the Sverdlosk data are representative. Figure 11.5 shows the same results, now compared with the results that would occur either if antibiotics were distributed more rapidly (the two rates were obtained by scaling plans from two different public health agencies) or if the antibiotics were distributed before the attack and taken in response to announcements made after the attack. This chart shows that distributing antibiotics at the higher rate in response to medical detection is as effective as starting distribution at the slower (but still high) rate in response to an environmental detection. To get a really effective response that holds fatalities to 1000 or so will require both the aggressive response to environmental sensors and rapid (or pre-attack) distribution. Figure 11.6 shows the same thing under the assumption of the incubation period derived from the U.S. data; in that case, because all the cases occurred within six days of exposure, one needs very rapid (or pre-attack) distribution of antibiotics coupled with an aggressive response to environmental detections in order to have any impact at all. The difference between Figures 11.5 and 11.6 is inherent to our limited understanding of the incubation of inhalation anthrax in humans. Decision makers and planners must

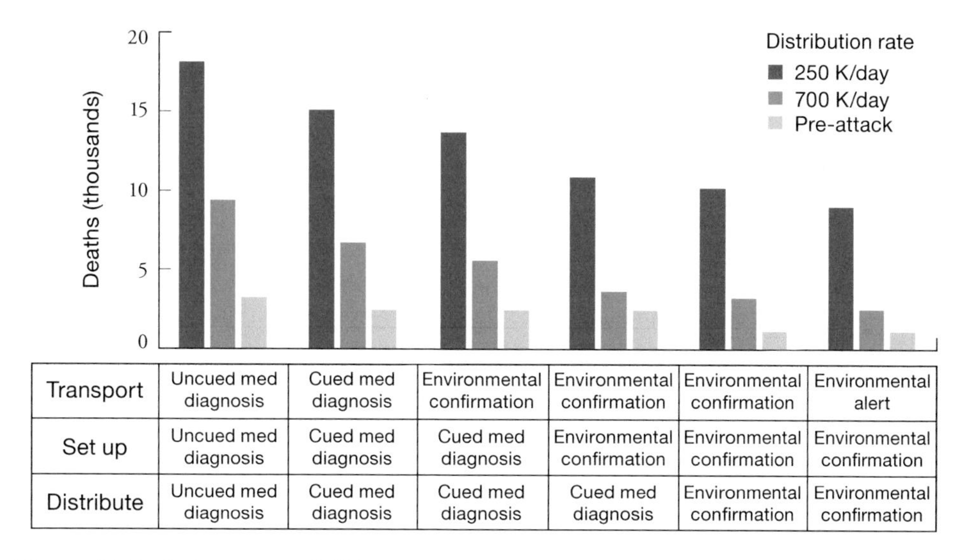

Transport	Uncued med diagnosis	Cued med diagnosis	Environmental confirmation	Environmental confirmation	Environmental confirmation	Environmental alert
Set up	Uncued med diagnosis	Cued med diagnosis	Cued med diagnosis	Environmental confirmation	Environmental confirmation	Environmental confirmation
Distribute	Uncued med diagnosis	Cued med diagnosis	Cued med diagnosis	Cued med diagnosis	Environmental confirmation	Environmental confirmation

Figure 11.5 Performance of different architectures given the Sverdlovsk incubation data.

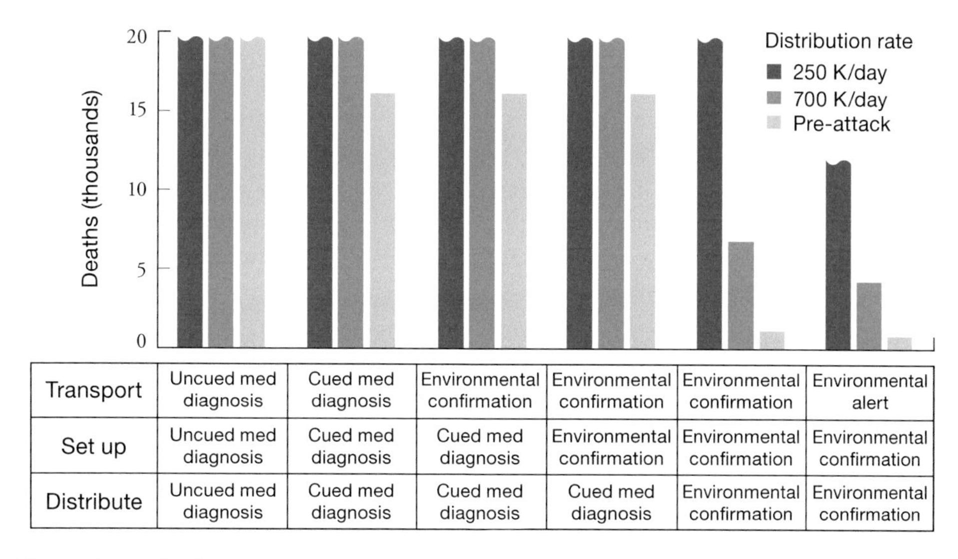

Transport	Uncued med diagnosis	Cued med diagnosis	Environmental confirmation	Environmental confirmation	Environmental confirmation	Environmental alert
Set up	Uncued med diagnosis	Cued med diagnosis	Cued med diagnosis	Environmental confirmation	Environmental confirmation	Environmental confirmation
Distribute	Uncued med diagnosis	Cued med diagnosis	Cued med diagnosis	Cued med diagnosis	Environmental confirmation	Environmental confirmation

Figure 11.6 Performance of different architectures given the anthrax mailing incubation data.

live with that uncertainty, and either mount the aggressive response or live with the knowledge that the response system may not be effective against certain attacks.

The models used in these calculations were straightforward and simple models of fixed delays for the various steps, uniform distribution rates, and so on. Nonetheless, we were able to draw several key insights from these calculations. The first is that the speed with which the antibiotics are distributed is at least as important as the aggressiveness with which the distribution is started and that both rapid distribution and an early start are needed, particularly if the tighter distribution of incubation periods observed in the anthrax mailings is representative of the attack; in that case, the time between the diagnosis of the index case and wide-scale illness is just too short. This insight suggested that the public health model, which is based on diagnosis of an index case, is not a useful way to respond to an incident of mass terrorism. The results also made clear that an effective and robust response would require both a prompt start based on environmental sensors and a rapid distribution. We also thought that if sensors could identify the portion of the population that had been exposed to the agent, a relatively small fraction of the populace of the metropolitan area, it would make possible very rapid distribution of antibiotics to those who really need them, although this response would be difficult both technically and politically.

The discussions that ensued when we discussed these results with public health officials were quite illuminating. They recognized the problem with waiting for medical diagnosis, but were still reluctant to respond significantly to environmental sensors. That reluctance was partly because the public health culture is one of responding after an index case, partly because many of the officials did not really believe in bioterrorism and also partly because they were concerned that even the best environmental tests did not distinguish between live and dead agents or between virulent and nonvirulent strains of the pathogens. The officials were concerned about panic that might result if even early steps, such as alerting physicians, became known. The anthrax attack attempted by the Aum Shinrikyo group in Japan, which failed at least in part because they used a nonpathogenic strain of the *Bacillus*, strengthened this concern.[1] The local officials we spoke with were also concerned about asking the National Strategic Stockpile for medication and supplies on the basis of data from environmental sensors; they feared that if the alert proved a false positive, they would lose credibility and not get help when they really needed it.

Distributing antibiotics before the attack was also considered problematic. The

1 P. Keim, K.L. Smith, C. Keys, H. Takahashi, T. Kurata, and A. Kaufmann, "Molecular Investigation of the Aum Shinrikyo Anthrax Release in Kameido, Japan," *Journal of Clinical Microbiology*, vol. 39, no. 12, December 2001, pp. 4566–4567.

officials were concerned about storage conditions and were also sure that many people would self-medicate with the antibiotics for a variety of illnesses for which the antibiotics probably would not be appropriate rather than hold on to them to be ready in case of attack. Finally, the idea of sorting the population of the affected area by exposure level was totally unacceptable to public health officials. First, they did not see a significant benefit to spending the time required to sort a population into exposure levels even if it were possible to do so reliably. Second, they knew it would be next to impossible to deal with people angry at not getting treatment and unwilling to believe they didn't need it.

The exchange was, I think, useful both for us and the public health officials, even though nothing concrete emerged immediately. The discussions did help bring home to them the message that terrorist attacks might not be just like epidemics. Although the officials were not yet prepared to change their plans, they did start thinking about how to respond. And we had useful discussions about ways of enhancing environmental sensing to address their concerns; for example, very rapid culture techniques that could quickly show that the detected agent was viable would be very valuable. The lessons we learned about the context of public health decision making helped inform future analysis. The demonstration of the importance of rapid distribution led to a consideration of means other than preemptive distribution to accomplish the goals; other options mostly centered on the U.S. Postal Service. The discussions also led to a number of tabletop exercises, which included both technologists and health officials, to develop relevant technology and plans. The problems identified in treating people after a mass attack also led to an increased emphasis on developing techniques to minimize exposure in the case of attacks and to a focus on important classes of attacks for which that can be usefully done, such as attacks in subway systems, where stopping trains limits the spread of the agent.

Of course, we had held some discussions with public health experts before we began the analysis, but these discussions were far less productive than the ones we held after doing some basic analysis. It was much easier to draw insight from the experts when we were talking about specific proposals for how to use environmental sensors and what they would mean, quantitatively, for the result. This was also my experience in doing systems analysis for military sponsors. When officers were asked to discuss how changes in technology would affect their missions in the abstract, it was very hard to get much from them; confronted with more concrete proposals for how to use technology to improve the way they are able to execute the mission, they are much more informative. It is always important to talk to the operators because the people who are doing the job in the field are able to provide better insight into the operational context that shapes what they can do and how effective it is than any collection of plans and documents can provide. What I have found most helpful is to read up and talk to people briefly before starting an analysis of a new problem, and then meet up as often as possible after developing concrete ideas and doing some analysis. I have always gotten more useful feedback when the proposals are concrete than when the questions are open-ended such as, "How might you use environmental sensors?" Even when what you need are open-ended answers or explorations, my experience is that you are more likely to get them when there is a specific concept on the table for discussion.

11.5 Summary

In some ways, systems analysis in a new area is not very different from analysis in an established area. It is important to step back and look at the big picture, to get the details right, and to dive in. And it remains true that much of the best analysis produces insight that, when viewed in hindsight, seems as if it should have been obvious from the beginning. It is particularly important in a new area to get calculating quickly, without waiting to learn all the required domain knowledge. In the first place, even if your initial analysis proves to be wrongheaded and of no direct value, you will more quickly learn what the real issues are by analyzing than by planning analysis approaches. Secondly, you will likely find that the domain expertise you need is not what you thought before you started. I have also found that the experts who have the needed domain knowledge are usually willing, or even eager, to teach you what they know. It is important to listen well to those who have this needed knowledge and operational experience so that you develop an understanding of the context in which they operate. Finally, good analysis is usually convincing even to an audience that is originally resistant; be patient in briefing the results.

About the Author

Michael P. Shatz is the leader of MIT Lincoln Laboratory's Systems Engineering Group, which began in 2009 as the Engineering Analysis and Testing Group under his leadership. The group performs analysis and environmental testing of the sensors and communication systems built by the Laboratory at both the component and system levels. He served in the Systems and Analysis Group from 1984 to 1998, where his work included analysis of threats to precision-guided munitions, infrared systems, and the survivability of air vehicles against a variety of conventional and unconventional threats. He also founded and led the Advanced Systems Concepts Group from 1998 through 2009, doing the work described in this chapter. He was a member of the Air Force Scientific Advisory Board from 1999 through 2003 and has served on a variety of additional studies for the DoD and intelligence community. He is a graduate of the Massachusetts Institute of Technology and earned a PhD degree in particle physics from Caltech in 1984.

12

Systems Analysis in the Air, Space, and Cyberspace Domains

Jack G. Fleischman

> Everything must be made as simple as possible. But not simpler.
>
> —Albert Einstein

> In theory there is no difference between theory and practice.
> In practice there is.
>
> —Yogi Berra

12.1 Introduction

I got into the systems analysis business in 1988 after graduating from the University of Pennsylvania with a PhD degree in high-energy physics. Like many physics graduates of that era, I faced the choice of starting the hard postdoc slog (in my case at CERN [the European Organization for Nuclear Research] in Switzerland) to a permanent academic position or getting a real job that took advantage of the broad technical education I had had the good fortune to obtain. Taking a hard-eyed look at my prospects, I decided it was time for a real job. Once that decision was made, accepting an offer to work at Lincoln Laboratory was an easy choice. The Laboratory's close connection to MIT gave it many of the positive qualities of a world-class academic environment, and its special position between government and industry made the work interesting and of enduring value.

Having chosen to work at Lincoln Laboratory, I was still uncertain how my background in high-energy physics would be relevant to the concerns and problems of the nation's defense. The defense world was *terra incognita* to me, as were the large community of defense analysts and the smaller community of systems analysts. All I brought to the job, beyond my education and research experience, was the general belief that physicists, and in particular high-energy physicists, were inherently "high-bandwidth" practitioners of the scientific method, and thus were not reluctant to dive head first into new technical areas, ask hard questions, and go where the story took them.

12.2 Early Systems Analysis Work

My first piece of good luck was landing in the Systems and Analysis Group, led then by one of our authors and a great analyst, Alan Bernard. One day soon after I had joined Lincoln Laboratory, Alan was walking down the hall, presumably gauging who was busy so that he could assign a problem that needed a quick-turn answer. Alan's theory was that if you want something done quickly, give it to the person who is the most busy. He passed my office, hesitated for a second, and then walked in to present me with the problem *du jour*.

In this case, the problem was right in my field of expertise and had floated down to my level in an interesting manner. The defense problem was one of detection of stealth aircraft. An enterprising, well-known astronomer had speculated that the cosmic rays raining down through the atmosphere could serve as the illuminator for a celestial radar. That is, any object moving through the atmosphere—say a stealthy aircraft—would perturb the flux of downwelling cosmic rays, and thus an appropriate high-energy particle detector might be able to exploit the resulting signal to a militarily significant extent. This astronomer had immediately submitted an abstract for an important upcoming conference, stating that his discovery would obviate the benefits from stealth. The abstract came to the attention of a particular senator's staff, who alerted the Air Force, and so on, until the problem fell into my lap.

Thus was my introduction to system-level analysis of unconventional defenses, a topic that I returned to several times during the formative years of my time at the Laboratory. It was relatively simple to use my knowledge of high-energy particle interactions, material effects on the propagation of these particles, a bit of geometry, and the machinery of optimal detection theory to get to a result. The upshot was that a detector operating with a filter precisely matched to the material properties of the air vehicle of interest, flying at a known altitude, could in principle be constructed. However, such a system would have to be square miles in area in order to operate with reasonable probabilities of detection and false alarm, and would exhibit extreme sensitivity to uncertainties in air vehicle material properties and mass distribution. Thus, the proposed system was highly impractical. With this kind of an introduction to the art of systems analysis, I was hooked forever.

Activities in my first 10 years at Lincoln Laboratory included analyses and field work in a variety of challenging areas: radar remote sensing of targets obscured by foliage; modeling and flight-testing of new air surveillance approaches; survivability assessments of fourth- and fifth-generation aircraft in threat environments of concern; and, of course, analyses of unconventional surveillance and defense concepts. The following war story of an interesting encounter illustrates an analysis of an unconventional concept.

War Story

In the early 1990s, a researcher at the U.S. Geological Survey (USGS) claimed to be able to track supersonic air vehicles by using the array of seismometers deployed throughout southern California. These seismometers were placed across the region to monitor seismic activity, but the researcher found that the shock wave generated by a supersonic air vehicle was sufficient to register on surface-mounted, as opposed to underground, seismometers, and that the time history of seismometer signals recorded across the array could be used to infer speed and location (in two dimensions). Using these data, the researcher successfully tracked several space shuttles coming in for landing at Edwards Air Force Base while they were passing over the Los Angeles basin at supersonic speeds.

While this bit of work was interesting, though perhaps not of great military significance, it got much more interesting when this same researcher claimed that he had seismometer data indicating the existence of a heretofore unacknowledged supersonic aircraft, which he assumed was the legendary *Aurora* spy plane. Specifically, there were several months of anomalous seismometer data indicating atmospheric shock waves, all occurring on Thursdays in the early morning. The researcher concluded that the air shocks were consistent with a hypersonic (in excess of Mach 5), high-altitude air vehicle.

As usual, such a claim attracted notice, and the request to investigate the validity and potential military utility of this technique rolled downhill until it reached my level. We got our hands on the seismometer data and did our own maximum-likelihood-based analyses. Try as we might, using exactly the same data as the USGS researcher, we always arrived at barely supersonic solutions for all the anomalous cases.

We even tried more advanced approaches, including incorporating acceleration, and hence wavefront caustics, into the trial solutions. After slogging through the resulting hundreds of terms and making suitable approximations, the best-fit solutions were still barely supersonic, with modest decelerations. In the spirit of systems analysis, we also searched for alternative explanations, ranging from atmospheric explosions (not consistent with the data) to UFOs (considered unlikely when the times of these anomalous events respected the switch from standard time to daylight-saving time).

Finally, we considered the hypothesis that the USGS researcher had simply made a mistake. This conjecture turned out to be exactly the case, and the mistake was a consequence of his making an inappropriate trigonometric approximation that was valid only if whatever was causing the atmospheric shocks was actually hypersonic, such as a reentering space shuttle.

Further investigation showed that the Navy had been conducting testing with F-14s off the coast of California at the same time that these anomalous events occurred, and that the testing involved supersonic flight that came close to land. The sonic booms were the result of some of the pilots trying to decelerate below the sound barrier before going feet dry and not taking into account the fact that sonic booms generated over the water would propagate for some distance over land before dissipating.

The dénouement of this story brings to mind a famous line from the movie *The Man Who Shot Liberty Valance*: "When the legend becomes fact, print the legend." In this case, even though the USGS researcher was presented with the true story, to this day the Internet is rife with contentions that these seismic events were proof of *Aurora's* existence, and the false story (i.e., the legend) is a staple of certain cable television documentaries.

By the end of this period, I was working closely with the Air Force Red Team, discussed by Aryeh Feder in Chapter 6, and particularly closely with Air Combat Command's Special Program Directorate. It was during my work with Air Combat Command that I fully realized both the power and the potential dangers of the systems analysis philosophy: one had to take great care in how one simplified complex air vehicle survivability problems to make them tractable (as the quotation from Einstein at the beginning of the chapter points out), and just as much care in how one articulated the results of this simplified analysis. Thus, I worked hard to prevent the results of simplified analysis from being taken as gospel and being directly converted into, say, requirements for the F-35 Joint Strike Fighter.

I was also going up the management ranks by the end of my first 10 years, and became leader of the Systems and Analysis Group in 2000. It was then that I truly understood an adage attributed to a former director of Lincoln Laboratory: if you hire the right people, train them, and mentor them, then you can leave the office by noon every workday. That was perhaps an overstatement, given the kind of hours Lincoln Laboratory people typically put in, but the sentiment was clear—the most successful people knew how to nurture creativity in others, put together effective teams, and keep those teams on track.

I took that sentiment to heart over the next 10 years, especially in 2006 when I initiated the Space Systems Analysis Group and when I worked with the Air Force Red Team to start the Air Force Space Red Team. A foundational precept of the Space Systems Analysis Group was to take the systems analysis philosophy and methodologies that had borne fruit in the air domain and set them loose to work their magic in the space domain.

In recent years, I have had occasion to apply these same methodologies to the cyber domain. In the following sections, I discuss the issues attendant to applying systems analysis to the space and cyber domains, and compare those flavors of systems analysis with the much more mature systems analyses done for the air domain.

12.3 Systems Analysis for Space and Cyber Domains

This section is divided into three parts. The first part focuses on the space domain and illustrates how problems of space control are amenable to the application of systems analysis by virtue of the similarity of problems in the space domain to those in the air domain, where systems analysis has been applied successfully for more than three decades. The second part is a top-level introduction to a basic concept critical to space control, space situational awareness. The third part briefly discusses the application of systems analysis for the cyber domain and underscores the relative immaturity of the application of the systems viewpoint to that domain. Because most serious discussions of the issues in the military space and cyber domains are classified, throughout this section I will stick to generalities rather than delving into specific technologies or programs.

12.3.1 Systems Analysis in the Space Domain

> Space is big. You just won't believe how vastly, hugely, mind-bogglingly big it is.
>
> —Douglas Adams, *The Hitchhiker's Guide to the Galaxy*

Military operations in the space domain are usually grouped under the rubric of space control, defined as the ability to maintain access to and use of the space domain. This domain is generally defined to be altitudes above where "the atmosphere has a major effect on the movement, maneuver, and employment" of forces.[1] For this discussion, we can use 100 km as a rough dividing line between the air and space domains. In the space domain, space objects are labeled by the orbital regime in which they reside: low Earth orbit (LEO, with orbital altitudes less than 2000 km), geosynchronous orbit (GEO, with altitudes within a band of ±200 km centered on 35,786 km) and medium Earth orbit (MEO, for altitudes between LEO and GEO). Space control involves conducting operations *to know*, including the aforementioned space situational awareness (SSA), and *to act*, generally grouped under either defensive space control (DSC) and offensive space control (OSC). Even the simple matter of defining these foundational terms, particularly deciding whether an action falls under DSC or OSC, can incite a semantic battle in a room full of space control practitioners. To me, such argument is a clear indication that foundational systems analysis is sorely needed.

As mentioned earlier, the space domain and the air domain have many elements in common and pose similar challenges to military and intelligence users. The differences in systems analysis for these two domains are driven mainly by the fundamental precepts that inform concepts of operation and employment, as well as the relative levels of maturity of the technologies and enabling systems in each domain.

In the air domain, when considering, for example, the defensive counter-air (DCA) mission, one starts thinking about the problem in terms of an end-to-end engagement (or "kill") chain; this central organizing construct is carried through from system design to operational employment. Steve Weiner used the kill chain construct in his chapter on missile defense, as did Aryeh Feder for a ground-to-air DCA mission in his chapter on red teaming; a version suitable for an air-to-air DCA mission is shown in Figure 12.1. A very similar, though less detailed, engagement chain is shown in Figure 12.2 for a space domain mission of defending an on-orbit asset in deep space. The DCA mission starts with early warning and surveillance aimed at locating potentially hostile targets in the contested airspace; the equivalent action in the space domain is space-object surveillance and identification, which constitute the discovery component of SSA. A key difference between air domain and space domain defensive problems is that the search volume in a realistic space control situation completely dwarfs that for a typical airborne scenario by many orders of magnitude. Countervailing that fact is that most of the objects in orbit around Earth follow predictable trajectories; significant perturbations from these trajectories are costly in fuel, time, and mission impact, and hence provide a strong discriminant between uninteresting and interesting objects.

---

1 *Counterair Operations*, Air Force Doctrine Document 3-01, 1 October 2008; Interim Change 2, 1 November 2011.

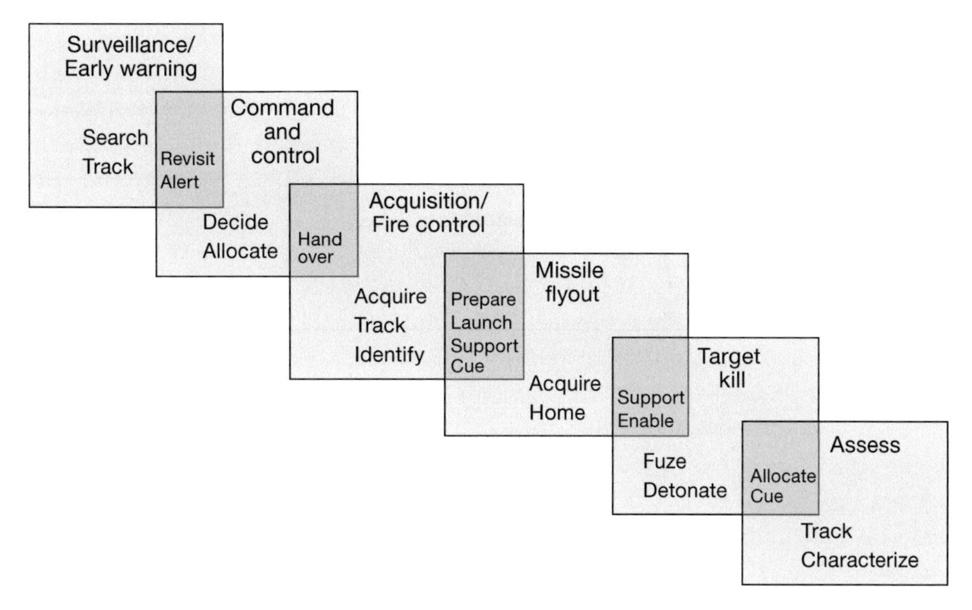

Figure 12.1 Notional defensive counter-air engagement chain.

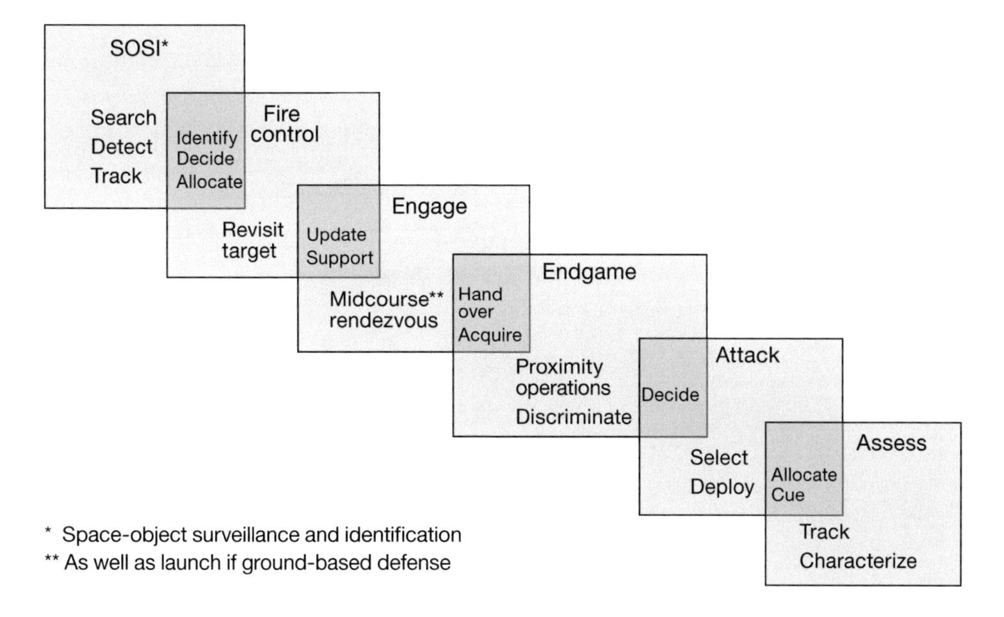

Figure 12.2 Notional defensive space control engagement chain, for the case of defense of a U.S. asset in deep space.

The DCA engagement chain depends critically on command and control (C^2), as well as communications. In the some 90 years of air power, C^2 has advanced, often by trial and error, such that in the 21st century, the operational level of air power projection has been refined by the U.S. Air Force to such an extent that it sometimes has the appearance of a turnkey operation. In contrast, U.S. space control has developed with a focus on SSA, but without the C^2 necessary to turn SAA data into action.

Further down the DCA engagement chain, a fire-control platform might receive a cue from a surveillance asset or collection of assets, obtain authority to generate lethal fires from C^2 elements, and perform the necessary actions to prepare a weapon for employment, including obtaining positive identification on the target. It is at this stage of the chain that the parallels between DCA and DSC break down, since up to the current day, the space environment has largely been a sanctuary for military and intelligence capabilities. Figure 12.3 illustrates the most important capabilities. Given the fact that space-faring nations have had the ability to fly with impunity in space up to the present day, few international agreements[2] bear on the issue of what actions are permissible in space. Similarly, no sentinel event has occurred that serves as a forcing function for the establishment of traditions or implicit understandings, as with early versions of the Law of the Sea, which has been continuously refined since the 17th century.

The similarity between operations in the two domains is reestablished by the need in both for battle damage assessment. The main differences between assessment in air and space are in the time scales and relative degrees of difficulty.

Much systems analysis for the space domain involves consideration of a future era in which the United States does have a robust set of options for mitigating threats to space-based services. The absence of fundamental precepts about admissible and inadmissible actions in space has multiple implications for systems analysis in the space domain. First, the very definition of a hostile action in space is hotly debated, so that systems analyses that depend on consideration of actions that follow a determination of hostile intent are immediately in definitional quicksand. The definition of what constitutes a space weapon is similarly on shaky ground. For instance, a state actor's right to conduct intelligence, surveillance, and reconnaissance in the air domain has well-established international norms; in the space domain, a state actor's observation of a space system not its own from less than some ill-defined range (for which there is no general international definition or understanding) is considered by some people

2 The international agreements most relevant to space control to which the United States is a signatory are the 1967 Outer Space Treaty (partially addressing interference and the use of weapons) and the 1972 Liability Convention (considering damage caused by space objects).

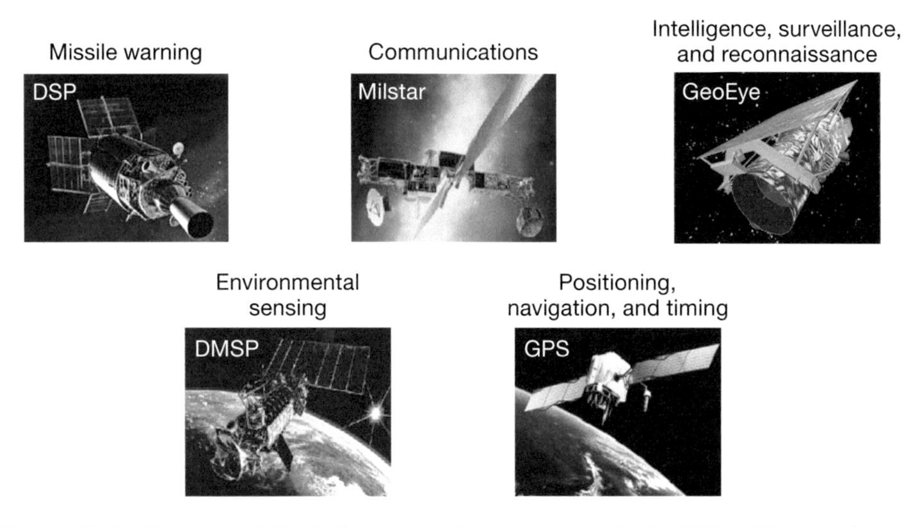

Figure 12.3 Elements of the U.S. space enterprise critical to the U.S. military and intelligence communities.

an act of war. There is also no common agreement on what is admissible in protecting a space system. Unlike in the air domain, where going on the offensive to protect oneself is considered normative behavior, in the space domain no clear dividing line exists between purely offensive and purely defensive actions. Given these gray areas, systems analysts in the space domain start their work somewhat shackled compared to workers in more mature domains, particularly with ever-shifting ground rules for what is admissible and not admissible in potential solutions.

That said, the fact that space control is relatively immature compared to air power projection is a reality that makes the application of systems analysis even more important. One way to see this need for comprehensive systems analysis is to deconstruct space control into its component parts, as shown in Figure 12.4. As can be seen in the figure, systems analysis can fruitfully contribute to understanding for a majority of these components.

12.3.2 A Quick Tour of U.S. Space Situational Awareness

As mentioned above, SSA is the critical enabling capability for space control across the engagement chain. The United States has the most well-developed suite of sensors for SSA in the world. For keeping custody of objects (live and dead satellites as well as debris) in the LEO regime (as well as objects in lower MEO), radar is the sensor of choice. In much of the rest of the world, space surveillance and tracking radars

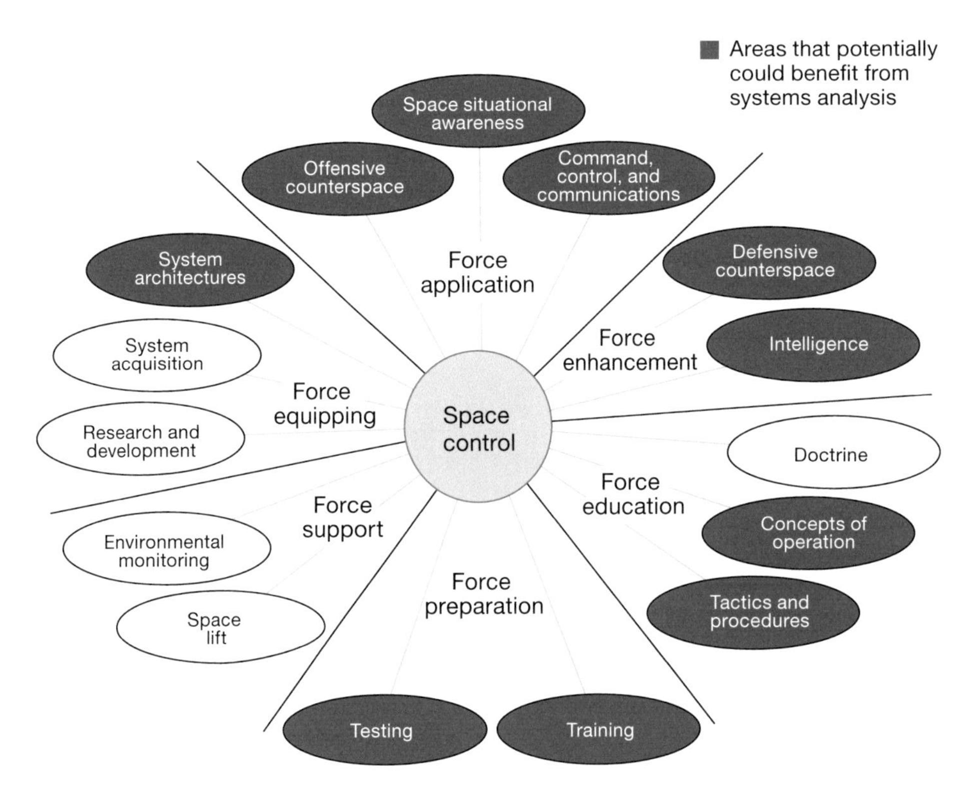

Figure 12.4 Elements of U.S. space control that benefit from the application of systems analysis.

exist only in near-peer nations; non-near-peer nations use optical telescopes for these orbital regimes. For upper MEO and deep-space orbits (i.e., highly elliptical orbits and GEO orbits), the world depends on optical systems, with the exception of a small number of highly capable space-tracking radar systems employed by the United States and near-peer nations. It should be noted that Lincoln Laboratory's Millstone radar and the Haystack Ultrawideband Satellite Imaging Radar (HUSIR) using its X-band capability, both located in Westford, Massachusetts, and the U.S. Army's ARPA Long-Range Tracking and Instrumentation Radar (ALTAIR) and Target Resolution and Discrimination Experiment (TRADEX) radar at Kwajalein Atoll (for both of which the Laboratory has a long history of providing technical innovation) are the only radars that contribute to U.S. deep-space SSA (along with the FPS-85 phased array radar at Eglin Air Force Base, Florida). In the optical realm, Lincoln Laboratory also has a long history of trailblazing and innovation, including building and operating the Space-Based Visible payload on the MSX satellite (a pathfinder for the operational SBSS satellite), as well as designing, constructing and commissioning the Space Surveillance Telescope, located in Socorro, New Mexico. The locations around the world of these systems are shown in Figure 12.5. The development and deployment history of U.S. SSA radar and optical systems is shown in Figure 12.6.

The fundamentals of surveillance and tracking system performance modeling discussed by Aryeh Feder in Chapter 6 and Dave Ebel in Chapter 9 are directly applicable to SSA modeling for space control. The key difference between modeling the flight of vehicles in the air and space domains is that space vehicles spend most of their time in predictable, near-Keplerian orbits, and spend a minority of their time executing maneuvers that involve expending precious propellant. Recall Dave Ebel's discussion of radar measurement error in Chapter 9; the same physics naturally holds true for radar detection and tracking of space objects. This fact can be seen in Figure 12.7, in which the elevation angle error as a function of single-pulse signal-to-noise ratio (SNR) is shown for the Cobra Dane phased array radar tracking the LEO-orbiting Starlette satellite (essentially a sphere with embedded corner reflectors used for geodesy and sensor calibration). The elevation angle error σ_θ decreases as the SNR increases and can be expressed as the sum in quadrature of the SNR-dependent error compo-

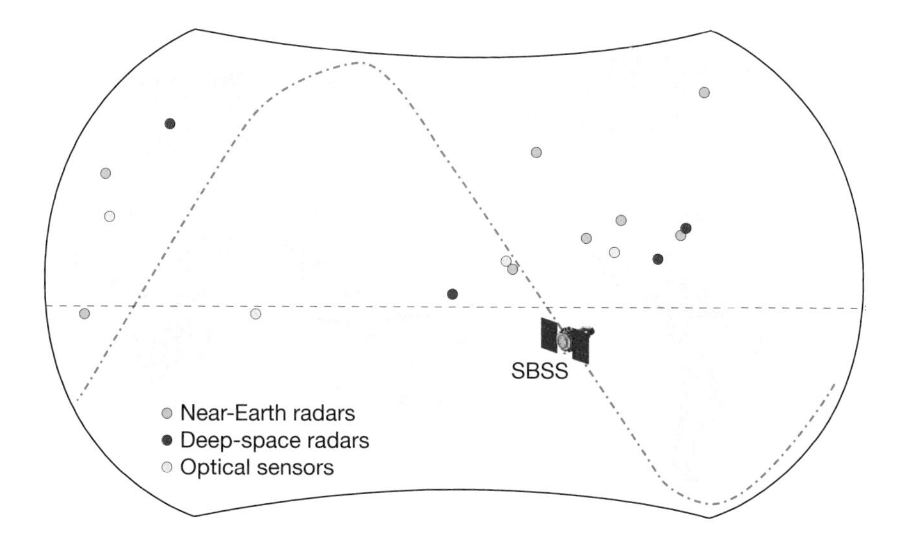

Figure 12.5 Locations of U.S. ground-based radar and optical systems. Also shown is the low Earth-orbiting Space-Based Space Surveillance (SBSS) satellite.

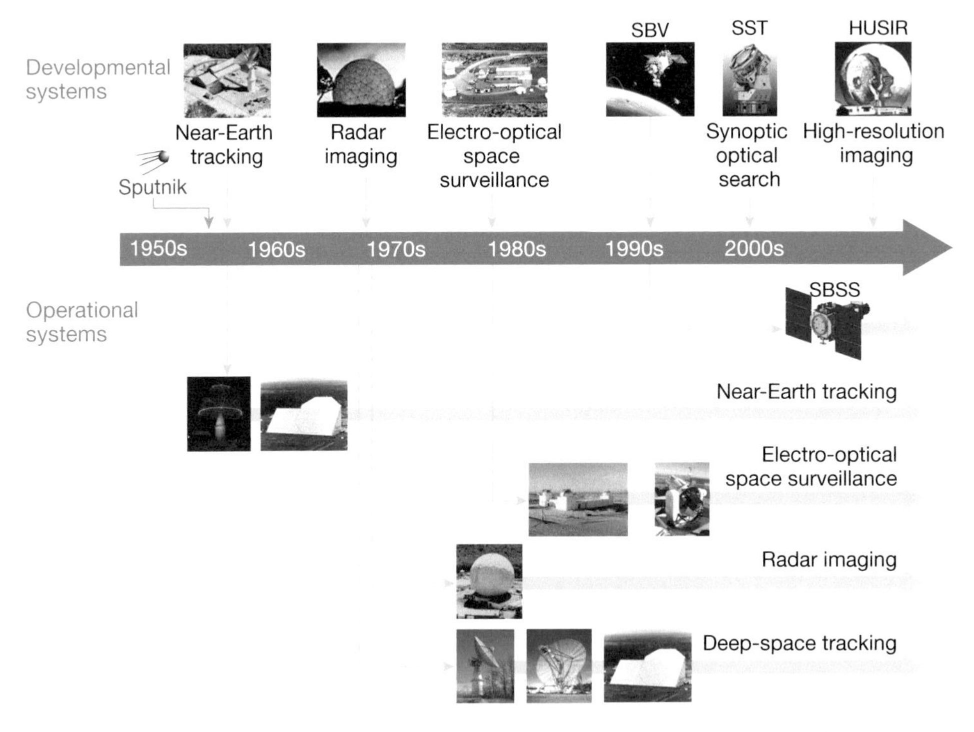

Figure 12.6 Timeline of developmental and operational U.S. SSA systems. SBV is the Space-Based Visible mission on the MSX satellite, SST is the Space Surveillance Telescope, and HUSIR is the Haystack Ultrawideband Satellite Imaging Radar.

nent and σ_{RIN}, the root-mean-square sum of all other elevation angle error components:

$$\sigma_\theta = \sqrt{\sigma_{RIN}^2 + \left(\frac{k \cdot \theta_{BW}}{\sqrt{SNR}}\right)^2}$$

Here θ_{BW} is the elevation beam width and k is a constant dependent on radar system characteristics. One can understand the systematics of optical system metric errors in a similarly fundamental manner.

While many elements of sensor operation and performance are common between the air and space domains, the specification of the state of motion of an object and the conversion of sensor measurements to state estimates (not discussed here) are notably different in the two domains. In the air domain, the position and velocity state of an air vehicle is usually specified in a Cartesian coordinate system attached to the Earth, for instance in Earth-centered/Earth-fixed coordinates.

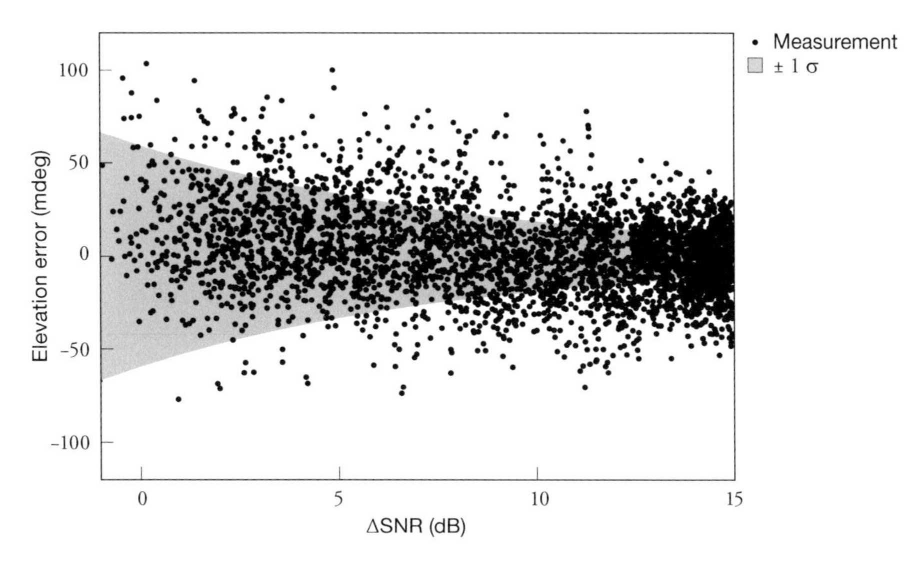

Figure 12.7 Measured elevation angle error as a function of single-pulse signal-to-noise ratio (SNR) for the Cobra Dane phased array radar tracking the Starlette satellite. The gray shaded region shows the ±1σ contours for the data, which decrease as the SNR increases.

For an object in the space domain, specification of its six-element state vector is a bit more involved. Figure 12.8 shows the coordinate system usually employed for specifying the motion of objects in Earth orbit. This specification starts with the definition of the ellipse traced out by an object in its orbital plane via the ellipse's semimajor axis a and eccentricity e; the eccentricity is given by $e = r_a - r_p / (r_a + r_p)$, where r_a (r_p) is the orbital radius at apogee (perigee) and $r_a + r_p = 2a$. The semimajor axis is usually reparameterized in terms of the so-called mean motion n, where $n = \sqrt{G(M + m)/a^3})$, with M and m the masses of the Earth and the orbiting object, respectively, and the units of n being revolutions per day. The specification continues with the definition of the orbital plane in terms of its three Euler angles, namely its right ascension of the ascending node (RAAN) Ω (which is the angle in Earth's equatorial plane between the location of the vernal equinox and the location at which the space object crosses the orbital plane in a south-to-north direction, also known as the ascending node), the inclination i of the orbital plane with respect to the equatorial plane, and the argument of perigee ω of the orbital plane (which is the angle from the ascending node to the point of perigee). The state vector is completed by specifying the angle ν (called the true anomaly) in the orbital plane between perigee and the angular position of the orbiting object at a particular time (termed the epoch). The

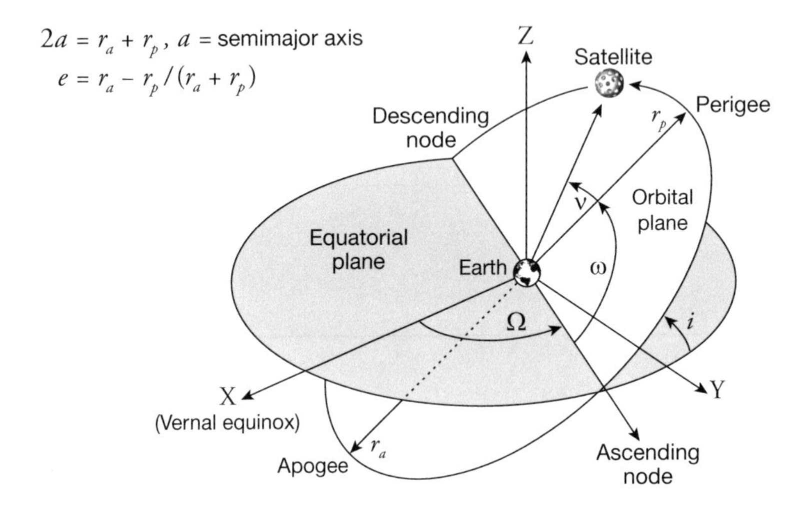

Figure 12.8 Coordinate system used for specifying the location of objects in Earth orbit. Variables are defined in the text.

true anomaly, in turn, can be obtained from the eccentricity and the mean anomaly (which is the angle ν calculated for an object in circular orbit about the Earth with radius equal to the semimajor axis of the orbiting object and mean motion equal to that of the orbiting object).

The U.S. Strategic Command's Joint Space Operations Center (JSpOC) disseminates a catalog of space-object orbits using the two-line element set (TLE) format. For example, a recent TLE for the Starlette satellite is

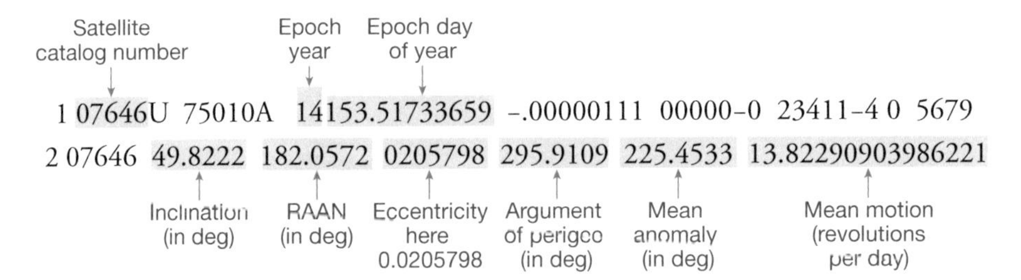

Orbit prediction with a TLE entails using what is termed a general perturbation model for propagating the position of a space object from epoch to the prediction time. General perturbation models incorporate the effects of Earth's mass distribution, atmospheric drag, and solar radiation pressure, as well as gravitational forces from the Moon, Sun, and the planets, into analytic expressions for the contributions to the acceleration experienced by an orbiting object.

For example, the gravitational potential from Earth's mass distribution can be written in spherical harmonics as

Latitude dependence

$$V = -\frac{GM_\oplus}{r}\left\{1 + \sum_{l=2}^{\infty}\left(\frac{R_\oplus}{r}\right)^l \left[J_l P_l(\sin\phi) + \sum_{m=1}^{l} P_{lm}(\sin\phi)\left(C_{lm}\cos(m\lambda) + S_{lm}\sin(m\lambda)\right)\right]\right\}$$

Earth's size, shape, density, and mass distribution (longitude dependence)

where G is Newton's gravitational constant, $M_\oplus$ is the mass of Earth, $R_\oplus$ is Earth's equatorial radius, r is the distance from the center of the Earth, the J_l are zonal coefficients that describe the variation in latitude of Earth's mass distribution, and the C_{lm} and S_{lm} are tesseral coefficients that describe the variation in longitude λ of Earth's mass distribution. Here the P_l are Legendre polynomials and the P_{lm} are associated Legendre functions (both depending on the latitude ϕ).

The contributions from selected terms for Earth's gravitational potential, as well as the contributions from other sources, are shown in Figure 12.9 in terms of the accelerations produced by these sources. One can see that Earth's oblateness, encoded in J_2, is the most important perturbation to the motion for orbital altitudes below GEO (the gravitational perturbations from the Moon and Sun becoming dominant for orbital altitudes higher than GEO). Atmospheric drag must be taken into account for the lowest orbits in the LEO regime; solar radiation pressure is the dominant nongravitational perturbation at GEO.

For the most accurate orbital prediction, the method of special perturbations is used. This method entails direct numerical integration of the equations of motion, using the force terms illustrated in Figure 12.9. In this case, orbital position at epoch is described by a JSpOC Vector Covariance Message, which contains not just the state vector but also the covariance of the state vector.

The accuracies obtained from a general perturbation model and the special perturbations method are shown in Figure 12.10, in which the measured tracking error (using laser ranging data as truth) for the Starlette satellite is shown as a function of time elapsed since the last observation. One can see that, in general, radar will produce a more accurate orbit prediction than an optical system will produce; this result is expected since radar measurements provide range and angle (and sometimes Doppler) information while optical measurements provide angle information only. One can also see that special perturbation solutions can result in orbital position errors that remain less than a few kilometers for several days and that can be stable for more than

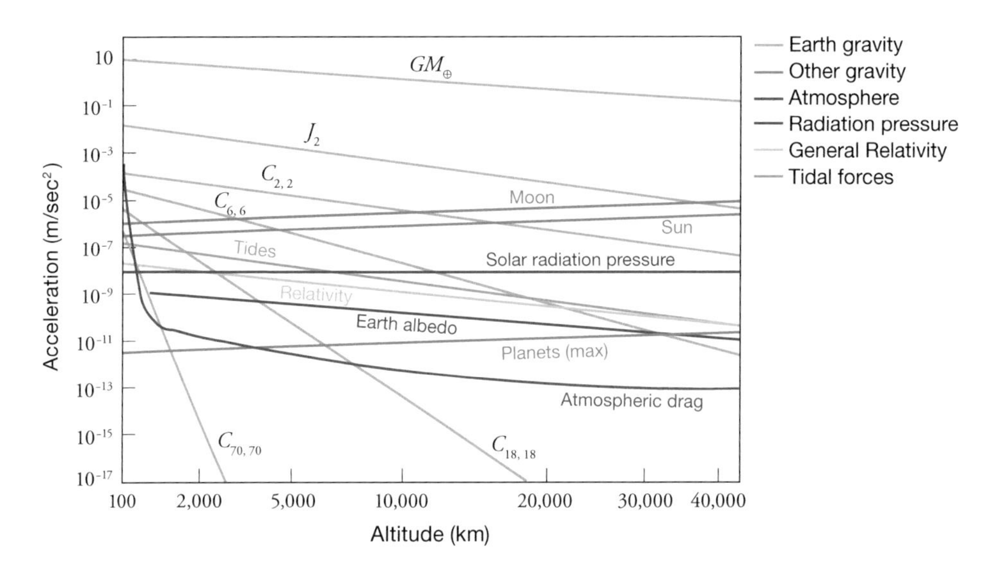

Figure 12.9 Accelerations produced by various sources on an Earth-orbiting object, as a function of orbital altitude.

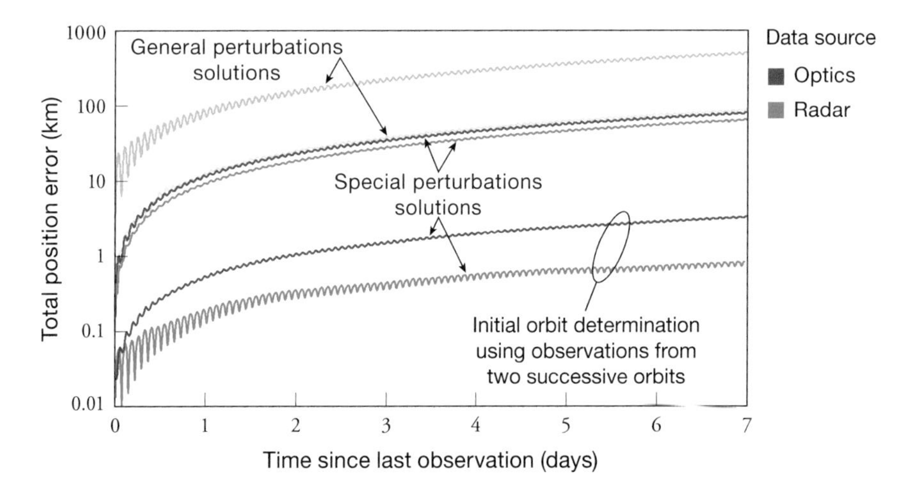

Figure 12.10 Total position error for the Starlette satellite as a function of time since last observation, using general and special perturbations. Blue curves indicate orbits propagated after initial orbit determination using radar data; red curves indicate orbits initially determined using optical data. The upper four curves use observations from a single orbital pass to estimate the orbital state vector; the lower two curves use observations from two successive orbits to estimate the state vector.

a week if observations from multiple orbital passes are used to estimate the initial state vector.

This rapid introduction to SSA for space control just scratches the surface of what is an operationally important (and technically rich) area of study. For instance, the relative stability of propagated orbits has relevance for defensive space control operations against potential threats.

12.3.3 Systems Analysis in the Cyber Domain

Given what I've written about systems analysis in the space domain, it should be no surprise that I feel just as strongly about the need for rigorous systems analysis in the cyber domain. Military operations in the cyber domain are more recent than those in the space domain, but state actors have seemingly been on a much steeper trajectory in terms of defining appropriate actions in cyberspace, assembling the *casus belli* for defensive and offensive cyber operations, and adapting well-understood concepts of command and control to the application of cyber capabilities. The driver of this rapid normalization is no doubt the short cycle time for innovation and the low barrier for entry in the cyber domain compared to entry for the space and air domains.

As discussed throughout this book, the engagement chain construct is often a useful tool for understanding the fundamentals of military operations; this is true as well in the cyber domain. Figure 12.11 shows a simplified engagement chain for defensive cyber operations, for instance, defense of a classified network.

Several elements of this engagement chain are particularly difficult for the cyber systems analyst. For instance, discovery, sensing, and assessment depend critically on persistent cyber situational awareness (CSA) because of the wide range of time scales over which an attack may occur—seconds to days for denial of service, seconds to months for exfiltration of data. Modeling CSA performance entails adequately capturing threat evolution over these disparate time scales. Similarly, assessing the effectiveness of threat mitigation involves capturing threat and counter-threat evolution by both the attacker and the defender.

These difficulties in applying systems analysis to the cyber domain should not be viewed as insurmountable. Similar complicating factors exist in the other domains, but are not as acute. Furthermore, developers of cyber capabilities tend to not adopt system viewpoints when considering their solution spaces. They favor the specific over the general, the present over the future, and fragile, single-point solutions over robust, integrated solutions. Thus, a cadre of disciplined cyber systems analysts is clearly needed; one hopes that such a community emerges in the next several years.

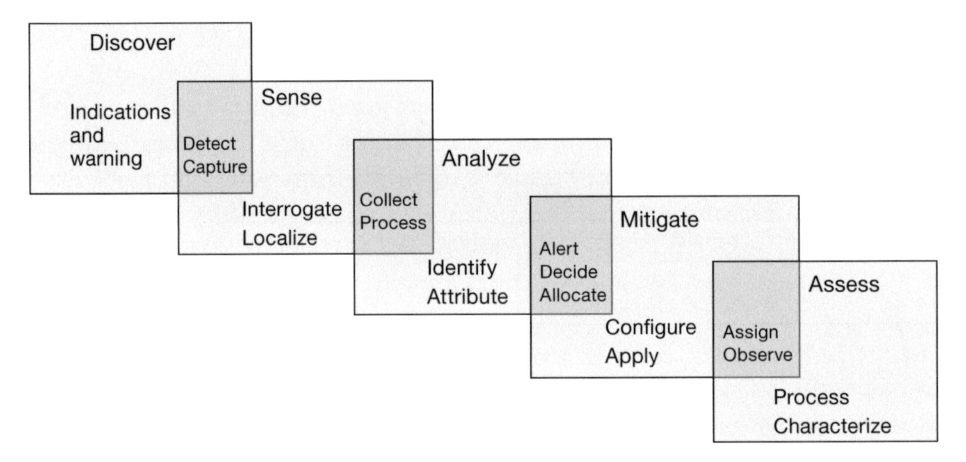

Figure 12.11 Notional defensive cyber operations engagement chain.

12.4 A Few Lessons Learned from 25 Years of Systems Analysis

I can't let this opportunity to get a few words down on paper end without relating some of the realities, at least to my mind, of the systems analysis business. My hard-won understanding of these lessons learned is a consequence of getting the occasional proverbial finger in the chest from senior military leaders, as well as from some of the true practitioners of the art who have mentored me throughout my career.

12.4.1 Don't Be Afraid to Recast the Problem

Often those who task you do not have the training or expertise to appropriately frame the problem of interest. Senior decision makers often do not have the time to think through what is worrying them or to understand in any detail the driving elements behind a problem or threat.

War Story

About a week before the Chinese successfully demonstrated their ability to engage a defunct weather satellite with a direct ascent anti-satellite (ASAT) system in 2007, we were asked by a senior Air Force leader to articulate which U.S. space systems would be at risk if the Chinese developed an operational capability based on their demonstration system. This decision maker was particularly concerned because he had been told that U.S. geosynchronous communication satellites and the Global Positioning System constellation would be "sitting ducks" for a potential Chinese ASAT. We immediately put together an assessment that gave the senior leader an easily digestible summary of the facts on the ground, indicating what was at risk and what was not at risk, not just from an operational version of the Chinese ASAT but also from other clear and present threats, as well as potential future threats. We further took the opportunity to lay out what we thought was the appropriate option space for U.S. actions in a future environment in which space was a generally contested domain. In the span of seven days of focused work, we helped to ensure that senior Air Force leadership had the best information that we could give them. This effort resulted in our being tasked to conduct a three-month study on options for responding to these emerging threats, and four months later to senior leadership's standing up a long-term analysis and testing team at Lincoln Laboratory dedicated to understanding the threats to U.S. space-dependent capabilities and assessing approaches for mitigating these threats.

12.4.2 Respect the Limitations of a Quick Analysis

The aphorism that Al Bernard mentioned in his chapter bears repeating: "a systems analyst will assume everything except responsibility." While this may sound overly harsh, as most systems analysts are trying to do the right thing, there is a kernel of truth to the statement. Systems studies are often tasked (and hence executed) in haste, especially if the person or organization asking for the analysis is in a short decision cycle or is trying to reach a decision milestone.

Given that a significant fraction of systems studies are conducted in a time crunch, compromises will be inevitable in the scope, depth, and generality of the analysis. Often the analysis that can be done quickly, while capturing some elements of a complex problem, may neglect important problem drivers. It is critically important that when this situation arises, the systems analyst attempts to bound the errors generated by simplifying the problem and does everything possible to ensure that the recipients of the systems analysis, both known and unknown, are aware of the limitations of the analysis.

Even for those systems studies that enjoy the benefit of adequate time, the systems analyst must guard against inappropriate simplification just for the sake of explication. By this, I mean analysts must always guard against going with a good story instead of facing up to a multiply connected, dynamic problem and managing its complexity and uncertainties. A good story has its place, but only if the problem has not been simplified to the degree that it has lost its relevance to the real world.

12.4.3 Be Vigilant Against the Inappropriate Use of Results

Most systems analysts with a few years in the business can reel off war stories about consumers of their work taking the analyses out of context or beyond the range of validity. Generally, the more powerful your analysis is, the greater the likelihood that others will want to adopt your work. The risk that this adoption of your work causes is that you can, and usually will, lose control of your message, even when the consumers taking your message and recasting it have the best of intentions. What invariably happens is that if consumers put their own wrappers around your work or use your work as input to their own analyses, the consumers will de-emphasize your findings and conclusions, and will emphasize their own interpretations of the situation instead. To make matters worse, consumers may also state or imply that you support their findings and conclusions since you were a contributing element in their study process.

Three courses of action may be taken to guard against inappropriate use of your systems analyses. The safest course is to resist releasing your results except to key decision makers, though this option will result in a reduction in your influence on and relevance to important technical and policy discussions. The second course of action is to broadly release your analyses, but only in briefing form with a live briefer attached to it. This approach can result in your spending a great deal of time away from home, but it will help minimize the chances of your work being misappropriated. The last course of action requires the most work: release your product in a fully documented final report. This option includes clearly stating the caveats pertaining to your analysis and making sure that each figure in your report can stand on its own (some consumers of your work will use the figures and ignore your prose). For complicated problems, the last option often takes a lot of thought, particularly in ensuring that you have selected the most incisive variables to plot. The virtue of this course of action is that a paper trail of your work will exist, making it difficult for consumers of your analyses to claim that you support their findings and conclusions.

12.4.4 Be Wary of Others' Agendas

Lenin wrote about the "useful idiot" in the political realm who unwittingly plays into the hands of extremist elements by taking actions that naïvely help extremist causes. In Washington defense policy and acquisition circles, the competition for resources and influence is usually a zero-sum game. Systems analyses that support the views or agendas of stakeholders in a particular area are often adopted or sometimes outright appropriated by partisans in order to gain a competitive advantage. Control of your product is critical to avoid serving as a useful idiot for an individual or organization that is much less concerned about the caveats and limitations of your systems analysis than you are. The three courses of action discussed in the previous section apply to this situation, but even greater vigilance must be exercised to prevent misuse of your analyses.

Exerting greater vigilance includes learning about your audience. Ask questions of informed people about the agendas and biases of the people you are expecting to brief. If your audience is mostly scientists and engineers, you can expect that respect for the truth will usually win out. If your audience is more nontechnical, with a focus on programs and policy, you should be aware that you may have people in your audience who may not be averse to using your work to further their aims. Your style of briefing should always be well matched to your audience, but in the case of an audience that includes those who may misuse your work, the words you choose should be carefully selected to clearly state the problem, articulate the caveats and limitations of your analysis, and clarify the generality and robustness of your results.

12.4.5 Keep Your Models Close and Your Understanding Closer

Just like the counter-air analysis community, the space control analysis community has its collections of "industry" codes that many analysts, particularly in industry and in program offices, take as gospel and use in an almost black-box manner. Many systems analysts eschew blind allegiance to these codes because those with expertise in relevant areas have deeper technical understanding of the systems and phenomena at issue than the creators of the industry codes. Industry codes are nonetheless part of a systems analyst's armamentarium. An analyst will often dissect them to understand the codes' limitations, simplifications, and outright errors, mainly to take on the inevitable questions that arise when there are differences in analysis output from the industry codes compared to the locally grown simulations that are usually employed.

As Aryeh Feder and Dave Ebel discuss in their chapters on red teaming, it is vitally important that the modeling to support a systems analysis is as closely coupled to laboratory and field measurements as possible. It is only when the modeling is subjected to experimental facts that the analyst becomes well aware of the applicability, limitations, and scalability of the analysis. With this philosophy, a natural hierarchy of engineering models, engagement-level tools, and scenario-based simulations arises, all supporting arrival at the so-called big picture illustrated in Figure 12.12.

A good systems analyst also follows the precept taught to graduate students in physics and other fields: always know (at least to an order of magnitude) the results of your analysis before you start an involved simulation effort. The famous Fermi problems, such as Fermi's oft-quoted estimation to within an order of magnitude of the number of piano tuners in Chicago, is an example of the kind of intellectual exercise a systems analyst should do before embarking on a detailed analysis. When one does embark on the detailed part of the analysis, one should know exactly what is in the simulation codes that are being used, and absent that, one should have an expert who does have this knowledge within arm's reach.

For the simulation code itself, above all demand transparency in its results; that is, always have a healthy distrust of what the code is spitting out, check the limits where analytical solutions can be derived, spot-check results for those situations in which analytical solutions do not exist, and always sanity check the results against intuition and the initial order-of-magnitude estimates. A necessary point of view is that a simulation code doesn't create knowledge, but merely collates the implications of the knowledge that the creator of the simulation code has put into it.

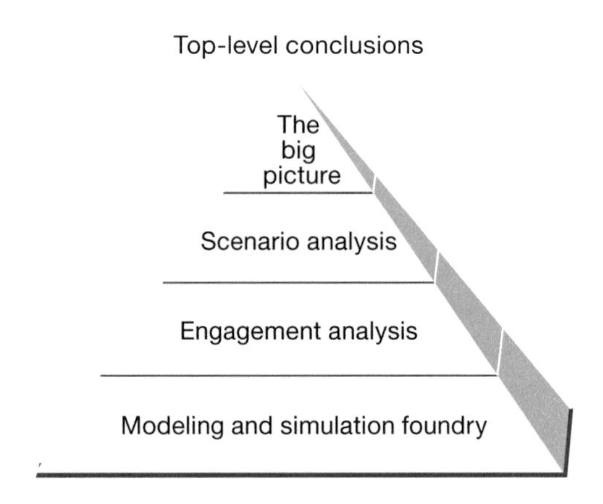

Figure 12.12 Hierarchy of models supporting systems analysis and the big picture in the space domain.

12.5 In Closing

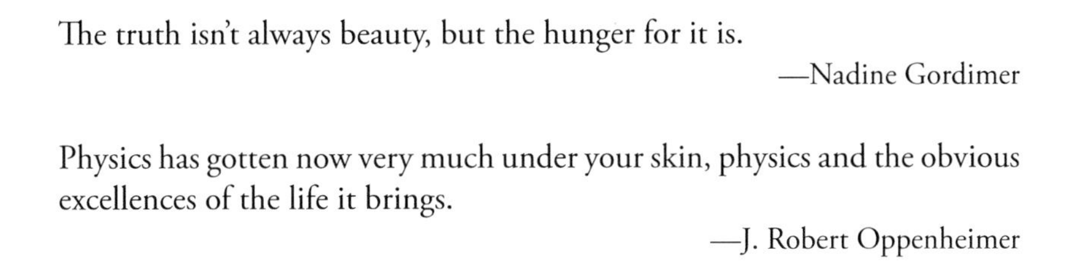

> The truth isn't always beauty, but the hunger for it is.
>
> —Nadine Gordimer

> Physics has gotten now very much under your skin, physics and the obvious excellences of the life it brings.
>
> —J. Robert Oppenheimer

I hope the words I've written earlier in this chapter are reminiscent of the two quotations above. Substitute *systems analysis* for *physics* in the second quotation and you'll understand the regard I have for systems analysis done well and for the best practitioners of the art. Systems analysis is not a replacement for in-depth investigation; it is a mode of thought that complements in-depth work that in the best of outcomes follows the path forged by the initial, often-simplified, system-level look. As Bill Delaney relates at the beginning of this book, being among the first explorers, or being the one who cuts through the fog of opinion and conjecture to get to the essential nugget of truth of an important problem can very much get under your skin.

About the Author

Jack G. Fleischman returned to Lincoln Laboratory in summer 2013 after serving since fall 2011 as the Chief Scientist of the Special Programs Directorate, Assistant Secretary of the Air Force for Acquisition. He currently is leader of the Computing and Analytics Group in the Cyber Security and Information Sciences Division. Before his tour with the Air Force, he was founder and leader of the Space Systems Analysis Group at the Laboratory. The group's mission is to assist the U.S. military and intelligence space communities in understanding and responding to emerging threats to space systems and services.

Before joining the Aerospace Division in 2005, Dr. Fleischman was leader of the Systems and Analysis Group in the Tactical Systems Technology Division. His work there focused on simulation and field testing associated with air vehicle survivability assessment, weapon system effectiveness, radar and infrared sensor performance evaluation, electronic countermeasure development, and mission modeling. Dr. Fleischman joined Lincoln Laboratory in 1988 after receiving a PhD degree in high-energy physics from the University of Pennsylvania. He has also been associated with the Air Force Scientific Advisory Board as a consultant and board member.

13

A Hundred Megabits per Second from Mars: A Systems Analysis Example

Don M. Boroson

13.1 Introduction

I joined Lincoln Laboratory in 1977, equipped with a PhD degree in electrical engineering from Princeton, where I had written a thesis using my skills in mathematics, statistics, control theory, and the like. I was really more of a generic applied mathematician than an electrical engineer. At the Laboratory in the Communications Division, I delved into the design and implementation of communication signaling and receiver architectures and analysis. I remember early on proudly writing a little internal memo analyzing, at the direction of my group leader, the detailed performance of some variant or other when a friendly, more senior fellow came by supportively asking me, "What does this REALLY mean?" and "How best should this influence our overall approach?" With that single interaction, I began my evolution into being a system designer and analyst.

Free-space optical communications, the area in which I have spent most of my time at the Lab since the mid-1980s, is truly a full-system design problem. There are myriad possibilities for the electro-optics parts and architectures, the telescopes, the mechanisms and controls, and the signaling itself. And each new problem comes with its unique constraints and difficulties—such as pointing knowledge of a spacecraft, atmospheric turbulence, background sunlight, crazy-fast desired data rates, low mass, incredibly long link distances, and so on. Selecting and then validating all the pieces to make the system buildable and successful in the face of those constraints is kind of an art. Each decision and selection influences all the other parts in a quite tangible way. One needs to continuously step back and see if the newest design addition invalidates some previous selection.

Although most of the other chapters in this book are concerned with defense systems analysis, it is not too much of a leap to include this chapter, which illustrates how we use such a system-wide view of a problem to "synthesize" a very new system.

13.2 NASA's Question

"Would it be possible to send data back from a satellite at Mars at 100 megabits per second using near-term optical technologies?" This was the question posed to some of us by a small visiting group from NASA (National Aeronautics and Space Administration). They had been told by others in the communications field that such a goal was far beyond what could reasonably be achieved at that time. So, the NASA team came to Lincoln Laboratory. They had heard that a team from the Laboratory had recently been the first to demonstrate high-data-rate long-distance laser communications (or as we call it, lasercom) from geosynchronous orbit to the ground in the GeoLITE program, and wondered if the demonstrated expertise found here could solve their own problem. Conversely, our team had heard little bits about this class of deep-space lasercom before and knew that it was being pursued mostly by a capable Jet Propulsion Laboratory (JPL) team. But we had never delved into this area of lasercom as it seemed quite different from the needs of our sponsors.

13.2.1 Background of Deep-Space Communications

Before I get into the problem itself and how we attacked it, I should probably give some background. So first are a few basic questions:

- How had NASA been doing deep-space communications all along?
- Why were they asking for this new data rate?
- What had prevented them from achieving this higher rate before?
- Why did they think laser communications would be able to solve the problem?

Let's start by understanding two basic tenets of communications, and digital communications in particular: (1) electromagnetic waves, i.e., radio or optical beams, spread out as they travel through space (or air, for that matter) because of what's known as diffraction; (2) the *rate* that a digital signal can be sent, i.e., the number of bits per second, is proportional to the amount of *power* that one terminal can deliver to the other.

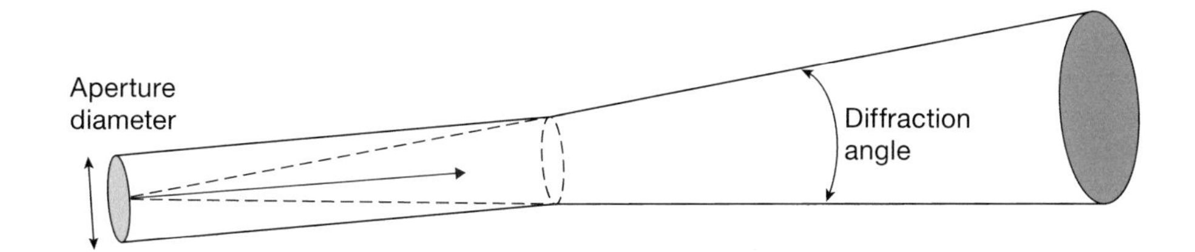

Figure 13.1 Diffraction of electromagnetic beams.

Figure 13.1 illustrates how a beam, again either radio or optical, travels. If the antenna, which for the optical signal is a telescope, is well designed, the beam starts out traveling in a straight path. Most people's vision of a laser beam in the lab is like this—the beam stays nice and compact. However, after a certain distance, all beams actually begin to spread out at a fixed *diffraction angle,* which is proportional to the wavelength of the electromagnetic wave divided by the diameter of the antenna. Because the wavelength of an optical signal is much shorter than that of a radio signal (as we'll see soon), its diffraction angle is typically much smaller than that of the radio signal, but it does spread out.

We can now see what happens. If we start out with, say, one watt filling the aperture, then at some point, the total energy one could capture with that same telescope would be, say, one-tenth of a watt. At some farther point, it would be one-hundredth, and so on. In fact, you can convince yourself that any receiver aperture would have its total received energy go down by a factor of R^2 if it were sent a factor of R farther away. (The diffraction is in two dimensions, of course.) This is the famous "R-squared loss" that is basic to communications and radar technologies. As the link's range gets longer and longer, a fixed-size receive aperture captures less and less of the transmitted power.

So now we can turn to the second basic tenet: achievable data rate is proportional to the received power. As we try to communicate farther and farther away with a single design, the data rate will have to go down—by the factor R^2.

We can summarize these concepts with a single equation, the so-called link equation, which takes into account the effects of the transmit and receive apertures in a symmetric way:

$$\text{Data rate} = E \cdot P_T \cdot A_T \cdot A_R \,/\, (\lambda \cdot R)^2$$

Here,

R = link range (in meters)

λ = wavelength of the carrier (in meters)

P_T = transmitter power (in watts)
A_T = area of the transmitting dish antenna (in meters^2)
A_R = area of the receiving dish antenna (in meters^2)
E = the efficiency of the receiver, in bits per second per received watt

This equation does not take into account other losses such as transmission through pieces of glass and through the atmosphere, but those losses could be easily multiplied in if needed. Otherwise, this formula is exact and is the one we use.

To give a better feeling for what all this means, let's take a look at some really long distances; in fact, let's look at the kind of distances that NASA is interested in. The chart in Figure 13.2 has distance along the *x*-axis and data rate along the *y*-axis. Because NASA deals with such great distances, we are forced to use a log scale on both axes.

Along the *x*-axis, we can see the following:

- Airplanes that can see each other are approximately 10s of kilometers apart.
- LEO (low Earth-orbiting) satellites are 100s of kilometers above the Earth's surface.
- GEO (geosynchronous) satellites are approximately 40,000 kilometers above the Earth.
- The Moon is 400,000 kilometers away from the Earth.
- The nearest planets are 10s to 100s of millions of kilometers away from the Earth.
- The distant planets are billions of kilometers away.

Along the *y*-axis are data rates from megabits per second to multiple gigabits per second.

In Figure 13.2, the diagonal line corresponds to the capability of a single communication system design. We see that if we had a system that could deliver 5 Gbps at geosynchronous distances, but then we flew one end of the link far away, the data rate would drop because the received power would be going down. At 100 × 40,000 kilometers away, that same system could deliver only 5 Gbps/100^2 or 500 Kbps. We should point out that the electronics to create 5 Gbps are very different from those that create 500 Kbps. So likely, the electronics to support systems along this line would have to vary. This curve only suggests what physics constrains us to. We engineers have to work out the rest of the details ourselves.

Figure 13.3 is a variant of the chart in Figure 13.2 but with several demarked regions: one with a system that can achieve approximately 5 Gbps if it were at

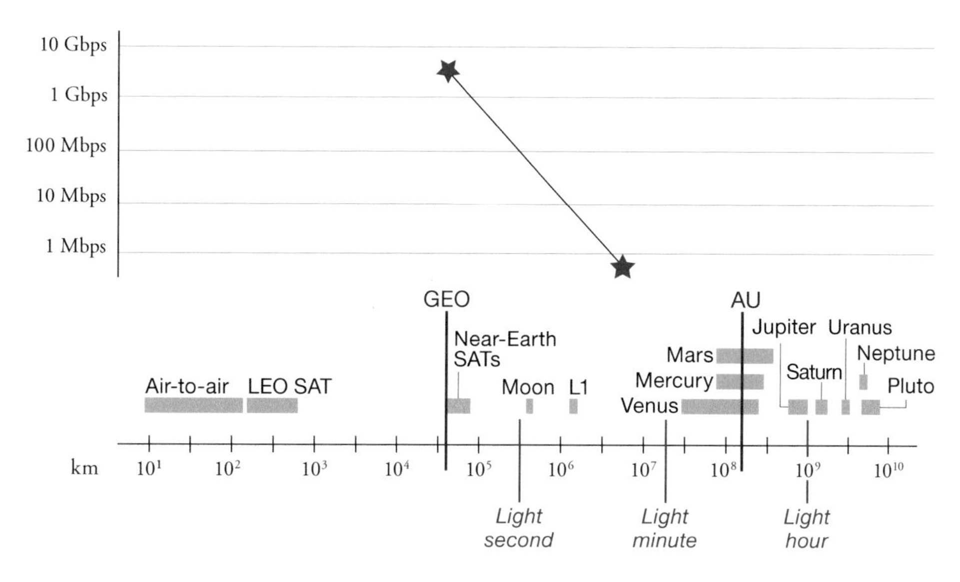

Figure 13.2 Constant difficulty systems through the Solar System.

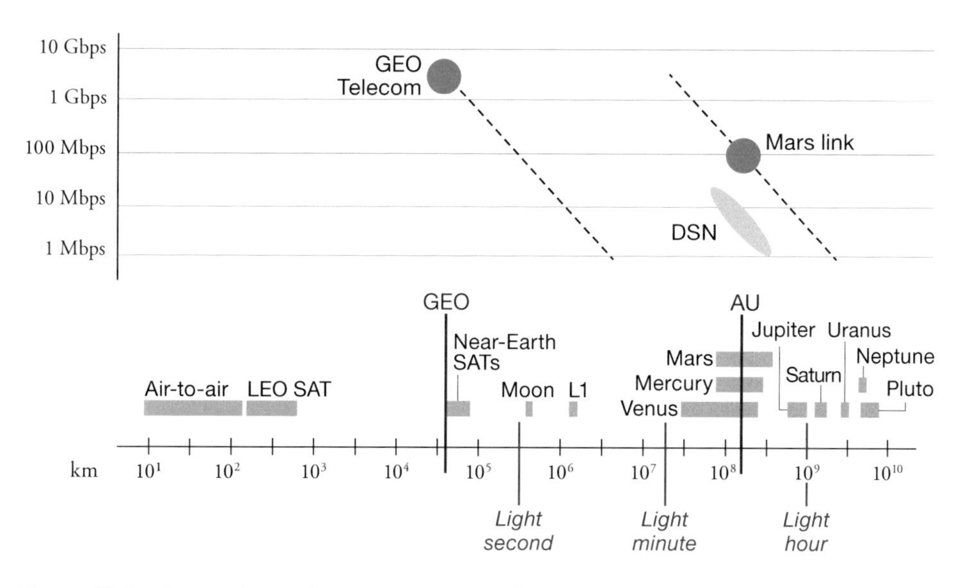

Figure 13.3 Comparison of near-term geosynchronous capabilities and the desired Mars capability, along with the current Deep Space Network radio capability.

geosynchronous distances, and one with a system that can achieve 100 Mbps if it were at Mars. (Because planets travel around the sun at different speeds, sometimes they are relatively close to each other and sometimes they are on opposite sides of the sun, at their maximum distances. We have arbitrarily chosen an average Martian distance for this comparison.) A system that can achieve 100 Mbps at Mars is the capability that NASA asked us to consider.

The third region in Figure 13.3 is labeled DSN. Since the 1960s, NASA had built up their very capable Deep Space Network (DSN). At each DSN site are some very large antennas for communicating with deep-space satellites using radio-frequency (RF) signals. Figure 13.4 shows sites at which some of NASA's 34-meter and 70-meter antennas are located. Because the Earth turns, any ground-based system servicing planetary missions needs to include several sites spaced around the world so that at least one of them is always in view of the satellite. NASA's DSN has as its main system three large sites in California, Spain, and Australia.

Even with these huge antennas paired with the best space terminals that today's spacecraft could feasibly carry, the DSN is only able to support up to a few megabits per second from Mars when it is at its greatest distance from Earth. Thus, we can immediately see why NASA would need to design a new system if they really want higher data rates to Mars. But why *do* they need these higher data rates?

Figure 13.4 The Deep Space Network. (Photographs courtesy of NASA.)

13.3 NASA's Needs

Scientists love Mars. As one of the closest and most Earth-like planets, it seems to have a lot to tell us about the Solar System in general and, in particular, it may be a mirror for sequences of geological events here on Earth. Therefore, NASA had, and continues to explore, plans for satellites to orbit Mars and send data back to Earth.

What if we wanted to make a high-resolution map of the entire surface of Mars for Earth-bound scientists, Martian probes, and possibly even astronauts? Let's say we wanted the map to have one-foot resolution. (This is a pretty reasonable request, as maps like that are "easily" made at the Earth and they are found to be very useful.) Some simple calculus concerning the area of a sphere tells us there would be 1.6×10^{15} points to measure and then transmit. That's a very big number. In fact, we can calculate that at 5 Mbps running 24 hours, seven days a week, it would take nine years to send only one bit from each point! Likely we'd need eight bits from each point; or, in fact, many more. The camera for such a measurement is already well within NASA's capability, but waiting close to 100 years to get the data back would be unacceptable, and so scientists have been scaling back their expectations for years. Now suppose we could send data back at a data rate 50 times higher than this. It would then take only about nine weeks for one bit from each point. Sending the entire map in high resolution would then be within the realm of possibility for a single mission.

Scientists would also like to measure many phenomena at Mars: temperatures, changes in surface features, radiation levels, wind speeds. They would like to collect various types of images of Mars—for example, optical, synthetic aperture radar, and hyperspectral images. Certainly, even video images would ultimately be required, especially once humans begin to make planetary trips.

We now see why NASA asked us to consider high data rates. Scientists would love to break through their data-return constraints at these great distances.

So why optical? You can see in the denominator of the link equation that a short wavelength helps a lot in delivering power and data. Let us now look at Figure 13.5, the electromagnetic spectrum. Frequencies of typical satellite communications signals are in the 1s to 10s of GHz. Thus, their wavelengths are in the 30-centimeter to 1-centimeter range. Optical signals used in the fiber telecommunications industry have wavelengths around 1.5 microns. Thus, they are shorter by a factor of more than 10,000. The diffraction loss, for a fixed transmit aperture size, is thus about 100 million times less (that is, 10^4 squared) for an optical beam than for a radio beam. We admit that there are many other factors to take into account in the link budget, but we see that optical beams have the potential to deliver much, much higher data rates. The operative word here, though, is *potential.*

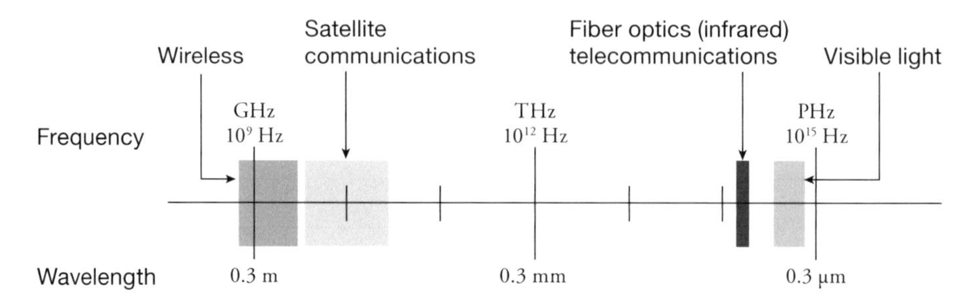

Figure 13.5 Useful communication bands in the electromagnetic spectrum.

So now back to the question at hand. By using near-term technology, would it be possible to engineer an optical system to actually achieve these high data rates from Mars? We told the visiting NASA team that we would spend some time thinking about it because it sounded "interesting." I guess that was a bit of an understatement.

13.4 First Impressions

I remember distinctly that after that first meeting, we went back to an office and did some simple link budget performance calculations on the board. A number of groups around the world had been working on 5–10 Gbps systems at the time, so we used as our baseline a 5 Gbps system at GEO distance. On the basis of that choice, it didn't take long for us to come up with a calculation similar to the one shown in Figure 13.3, which said that we would need to improve a present-day design by a factor of about 560,000! That is about 57.5 dB, for those who speak that language.

Our strong opinion at first was, "Wow, an improvement of half a million! That's crazy! This can't be done!" Engineers usually gulp when you ask for 3 dB improvement. A request for 57 dB improvement is just too funny.

13.4.1 Observations about General Engineering Design Approaches

Let's stop for a second, and think about how engineering is done. In one high-level view, there are two large classes of engineering design:

1. Try to engineer all the parts of a new system just *a little bit better* than with previous efforts, and try to squeeze out the desired improvement. This may be in performance, cost, complexity, ease of use, and so on. Trying to do anything

other than such an evolutionary approach usually means the program will be high risk.

2. Think "outside the box" in every facet of the system. The goals might not be otherwise achievable with standard approaches. Then hang onto your hats.

It became pretty clear to us right away that finding a performance improvement factor of half a million required the second approach in a number of the design elements of our proposed system. Our small group's approach to this outside-the-box need was to first understand the natural and theoretical limitations that bound the problem. Here the questions to ask were

- What is the largest reasonable power that we could ask of a satellite-based laser transmitter?
- What are the drivers on architecting very large ground receive apertures that have a chance of achieving the link efficiency proposed above, especially in the face of atmospheric turbulence?
- What is the highest theoretical efficiency achievable by transmitting and receiving laser beams?
- What are the drivers on stabilizing telescopes and beams on a satellite in deep space?

These are the basics that we kept in mind as we thought about each facet in turn.

13.4.2 Back to the Problem

Given these guidelines, we started to pick the problem apart from first principles:

- Perhaps we will be able to improve on our now-five-year-old GEO-class laser transmitter power by a factor of ten.
- Perhaps we could triple the diameter of our GEO satellite aperture, at least if it were not gimbaled (giving a gain factor of nine).
- We could, like deep-space radio engineers, greatly increase the diameter of the Earth-based receiver by a factor of, say, 25, giving a gain of 625.
- These improvements and changes would add up to about 56,000.

We were still a factor of 10 short, and we hadn't really done any of the (likely difficult) engineering to effect any of these improvements we've predicted. Someday in the future, transmitter power and aperture diameters will certainly grow. But in the near term, even these values would take a goodly amount of development. So to achieve the goal, we needed to do two main things: come up with feasible methods to achieve even these values and then find the missing factor of 10 (10 dB)!

13.5 First, the More Straightforward Elements

With regard to transmitter power, technology moves right along, especially when it's a technology of use to multiple industries (e.g., lasers are used in printing and welding, not to mention telecom), and so building an updated optical space transmitter with somewhat higher power should likely be relatively straightforward. Of course, the transmitter development will likely run into the usual "gotchas" of space-qualified parts, thermal design, etc. But it's at least feasible.

The second basic element of the system is the space optical aperture. Mechanical and thermal considerations for satellite telescopes will always need to be dealt with, but people have been flying pretty large telescopes for years, so the only real difficulty would be if mass were *really* constrained. (You can always make a somewhat hard problem *very* hard by adding more and more constraints.)

If we ask for the telescope to be gimbaled, we know it will require a lot of engineering care. So let's try to engineer our deep-space satellite terminal without a large gimbal. We'll depend on the satellite to coarsely aim our telescope. Deep-space RF antennas are pointed by the spacecraft, so this requirement is not too great a burden. Even without a large gimbal, however, we know that there will still need to be some mechanisms (or electro-optics tricks) to point and stabilize the very narrow transmit beam because the spacecraft can only assist with pointing to an extent. These mechanisms may require some clever designs. As with the transmitter, though, engineers have been pretty clever, and we felt there would be mechanisms that could be employed to achieve the required performance. Exactly how we would control those mechanisms, though, would still take some system engineering. We will return to that later.

So, what tasks are remaining?

- Design a very large, not ridiculously expensive ground receive telescope.
- Design a ground detector and receiver that achieve the theoretically best performance while sitting behind this huge telescope, all the while operating in the face of turbulence.

- Solve the space terminal's pointing control problem.

To understand some of these trades, we need another digression.

13.5.1 Digression—This Time about Atmospheric Turbulence

Laser beams love the vacuum of space. Nothing bothers them and they happily propagate and diffract exactly as the theory tells us they will. Propagating through the atmosphere, however, is a different matter.

When we think of a laser beam, we usually think of it as a narrow tube of light energy. We can also think of it as the electromagnetic wave, with each wavefront, one behind another, filling that column of energy. If the wavefront is flat, then we get *diffraction-limited* performance. However, if the wavefront is wavy or distorted, then the beam spreads out faster than we showed in Figure 13.1.

Unfortunately, small variations in the heat and thickness of the air in the atmosphere lead to such wavy wavefronts (Figure 13.6). If we transmit a diffraction-limited beam up through the atmosphere, it starts to accumulate these wavy wavefronts, and the beam begins to spread out a little bit. Once the beam hits the vacuum of space, there are no more distortions, but the damage has been done. Because of the now wavy wavefronts, the beam spreads out through the vacuum in a random way, and a far-away terminal will almost certainly see much less energy. It might even see zero energy if the beam happens to form the detail of a temporary null in that direction.

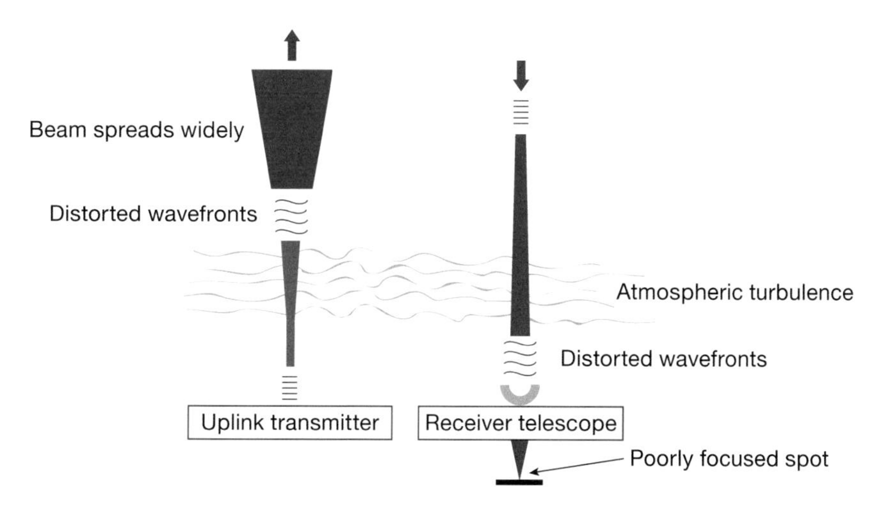

Figure 13.6 Atmospheric turbulence effects on uplinks and downlinks.

Downlink beams see something different. They have propagated from the far-away terminal very long distances through the vacuum of space according to the diffraction theory, and only at the end of their travel do they see the distortions of the atmosphere. The beam starts to spread a bit through the atmosphere, but it does not have a chance to spread very much because the atmosphere is only tens of kilometers thick. However, the wavefronts have become wavy. What that means is that a telescope does a good job catching the beam, but cannot focus it to a tiny spot. The focal plane of the telescope will see the dot grow and shrink and dance around.

So, we have two phenomena to cope with for our deep-space problem:

1. It is difficult to send high power to the spacecraft. How can we make the spacecraft work at all?

2. It is difficult to focus the downlink received beam to a tight spot, especially from a large telescope. How can we catch an adequate amount of light with a relatively inexpensive collector and then focus it onto detectors that are fast enough to support the changes in the signal caused by the high-speed data modulation? (Usually, the highest-speed detectors are particularly small.)

13.6 First Invention: Hybrid Tracking System

Optical beams are very, very narrow. Of course, that is why they are capable of delivering a lot of energy. However, pointing them to exactly the right place and holding them steady is a very tough job.

We invoked earlier that engineers would be able to create whatever mechanisms or devices we need to impart the steering and pointing corrections. What is *hard* is to know where they should be pointed or in which direction to make the constant pointing corrections. Traditionally, lasercom engineers have stolen a fraction (say, 10%) of the incoming light that was sent mostly for the communication function, and then sent it to a spatial sensor. The sensor can see when the spacecraft has moved because the focused beam's spot moves on the sensor. That motion can be turned into a correction direction signal that is usually sent to a little mirror behind the telescope that moves the beam back to the center of the sensor. If we send our outgoing beam off this mirror, then, by reciprocity, it continues to point right back at where the uplink beam came from! There are target-leading details, called point-ahead, that we will not consider here. Suffice it to say that this is another tricky problem that can be engineered with some cleverness and a lot of care.

The "gotcha" in this concept is that it is based on the premise that there is *enough* received light to make the measurement. It turns out that the faster the corrections need to be made, the more received power one needs. Unfortunately, even on a relatively "quiet" spacecraft, there are audio-frequency kinds of vibrations that need to be corrected.

Guess what? Unless one works very hard to transmit huge amounts of uplink power, there is likely not enough signal reaching the deep-space terminal for performing this tracking measurement. (We assume here that it would not be worth the huge cost to send a high *uplink* data rate to Mars—at least not until there are astronauts involved.)

What to do? Others in the field had considered looking back at the illuminated Earth in order to use its image as the reference signal for pointing and tracking. Unfortunately, it was found that it is extremely difficult to use the Earth's variably illuminated crescent shape to deduce exactly where to point the very narrow downlink. From Mars, the beam spot on the Earth is expected to be only a few percent of the Earth's diameter. Star trackers could probably give the pointing accuracy that we needed, but adequately bright stars do not usually fall in the field of regard of our lasercom telescope. (Perhaps someday someone will figure out how to use an off-axis star sensor to point the lasercom beam.)

As we tried to crack this nut, a few people on our team started to realize that we can break down the pointing and tracking problem into different frequency ranges. A transmitted weak uplink beam is probably a perfect reference for the absolute pointing, but not bright enough for wideband tracking. The Earth's image is not good for pointing, but might be bright enough for mid-frequency tracking corrections (when the orientation is such that the Earth is illuminated).

Finally, a member of the team invoked a specific type of inertial reference unit (IRU) device, first invented at Draper Laboratory, to deal with the high-frequency vibrations. IRUs have been widely used in military and civilian applications. In our application, the device was a small optical platform connected to the spacecraft by using a set of springs and pistons (Figure 13.7). The device was able to point a small, low-power beam very stably by using special inertial sensors that are mounted on the IRU itself. If we shine that laser into our telescope, it looks like a very bright star in our field of view, and we can track it with our traditional steering mirror method.

The complete tracking system, then, combines the three frequency ranges—uplink beacon at low frequencies, Earth image at medium frequencies, and the IRU optical signal at high frequencies—and points and stabilizes the duplex lasercom beam adequately to support the link even at very long distances (Figure 13.8). By dividing up the control frequency range, the team essentially solved the problem of deep-space beam pointing and stabilization. Of course, it was more complex and took a fair amount of software, but when you are dealing with problems constrained by first-principles physics, you are happy to find such solutions.

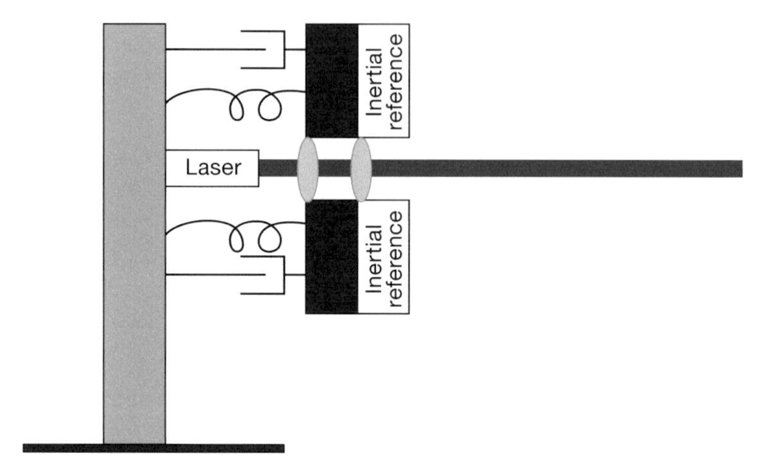

Figure 13.7 The inertial reference unit. The body has inertial references on it. They provide the feedback to the actuators pushing on the spring-mounted body. The laser beam is thus stabilized.

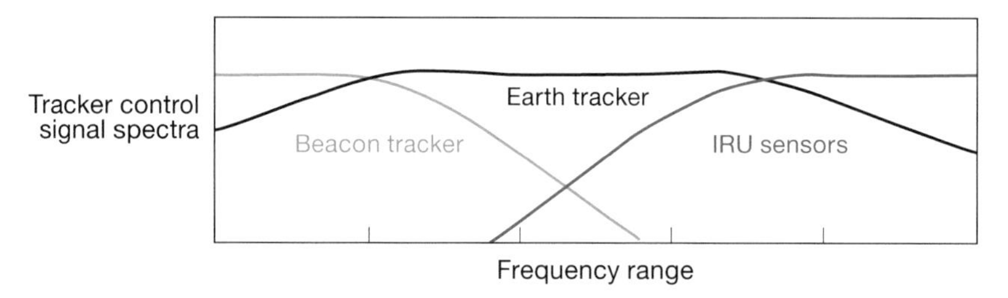

Figure 13.8 The hybrid tracking approach.

13.7 Final Problem—Telescope, Detectors, Efficiency, and the Signal

So from our initial cut at the problem, we still needed to find a way to

- Create a not-too-expensive, Earth-based collector of 5–10-meter diameter,
- Have it couple its light onto some kind of fast detector, even in the face of turbulence,
- Try to do this and be a factor of 10 more efficient than anyone had been before.

Piece of cake! We started with the efficiency question.

13.7.1 What Is the Highest Possible Efficiency in Any Lasercom Receiver?

The modern era of lasercom at Lincoln Laboratory began around 1980 when the newly formed team tried to achieve relatively high efficiency by using the coherent technique known as heterodyne. In fact, this is the technique that some radio engineers use for their systems. In those days before optical fiber, the beam from a semiconductor local oscillator laser at the receiver was carefully overlaid spatially onto the incoming beam, and the total signal was measured on a high-speed detector. This approach generates the beat tone between the two frequencies, and one can demodulate the data by using standard radio-electronics techniques. Such a system achieved efficiencies of about one bit for received energy corresponding to about 10–20 photons. Such performance was excellent in its day, and our group received a lot of attention (and sponsorship) for this work. Other potential coherent receivers, in which the local oscillator was actually phase-matched to the incoming signal in the process called homodyne, were extremely difficult to achieve with the lasers and optics of the day. As it turns out, both these types of receivers have since become feasible with the advent of narrow-linewidth lasers and single-mode fibers, as well as other components from the telecom industry.

Because we had always been interested in limits of efficiency, along the way we had performed a rather complete survey of all the possible methods of modulating and receiving communication laser beams. It was found that by using the Shannon's channel capacity metric, we were able to compare all possible performances, even without specifying the details of the implementation methods. What popped out to us was that a seemingly simpler approach, in which a particularly low-noise kind of detector called a noiseless photon counter directly measures the incoming light energy, had the tantalizing property of being even more efficient than homodyne. Furthermore, with modern error-correction codes, we predicted that we might be so efficient as to create our still-needed factor of 10!

13.7.2 On Noiseless Photon-Counting Receivers

As early as the 1960s, physicists had deduced that low-duty-cycle pulsed signals were optimum for photon-counting communications. In fact, a seminal experiment performed in a JPL lab in the early 1980s had successfully demonstrated the possibility of high-efficiency photon-counting lasercom.

The approach we use is called pulse position modulation (PPM), and we have sketched it out in Figure 13.9. If we take the original bit sequence and divide it up into symbols made up of K bits, we see that we can consider the stream as a sequence of K-bit words. Each such binary word corresponds to a number between 0 and $M = 2^K - 1$.

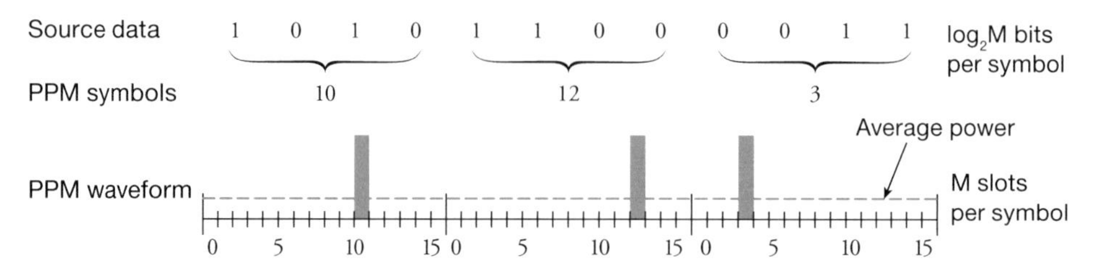

Figure 13.9 Pulse position modulation (PPM) and photon-counting.

So if we divide the time that the original *K* bits took into *M* new, faster timeslots, we see that we can transmit this sequence by sending a single pulse per word. (Of course, the transmitter to create such a signal needs to store up its energy and then send it all out in one faster, high-peak-power pulse.) We know that, after a large link loss caused by the transmitted beam diffraction, there might be only a single photon remaining in the pulse. However, we can also see that detection of that single photon is all we need in order to return to us all *K* original bits! So, except for possible lost pulses and the effects of noise, we see that it might be feasible to send up to *K* bits per received photon. It sounds like a swindle, but we are paying a rather large price in bandwidth and speed of electronics to pull this off. In the laser business, we do have lots of bandwidth. Remember, 20 years ago we were happy with 20 photons for each bit. This new system can potentially provide multiple bits for each photon.

Of course, we do have to deal with those lost pulses and noise, so error-correction coding is mandatory in the most efficient photon-counting links. When the average rate of arrival of photons is, say, one per pulse, Poisson statistics tell us that there is a $1/e = 37\%$ chance that none will be detected. Thus, the code needs to deal with a sizeable fraction of such "erasures" as well as the effects of false counts resulting from extraneous light sources. Luckily, there is much literature on possible codes, and we ended up choosing a code and decoder architecture originally developed by Caltech and JPL.

Because PPM creates time slots quite a bit faster than the original bits and the error-correction code adds redundancy bits that further increase the slot rate, the photon detector needs to discern photon arrival times at a rather fast rate. For example, for $K = 8$, $M = 256$, and a rate ½ code, we see that the time slots come $2*256/8 = 64$ times faster than the original source bits. At 100 Mbps source rate, this rate would require 6.4 billion time slots per second. Similarly, at $K = 4$, $M = 16$, the slots would come only $2*16/4 = 8$ times faster than the source bits, but it turns out that $M = 16$ is not as photon-efficient as $M = 256$. There is thus a basic trade-off between efficiency (the larger K and M, the more efficient) and required detector speed (proportional to M/K) for any data rate.

Could we build a transmitter that could generate such a signal? Luckily, the answer was "yes." Using fiber telecom components, our team had for some years been building high-data-rate transmitters based on high-speed telecom modulators and doped-fiber optical amplifiers. It turned out that, if the data rate were high enough, we could use these very transmitters to create the PPM signals. It was only if we forced the system to run slowly that we would need to look elsewhere. Here was a case where forcing ourselves to solve the high-data-rate Mars link problem allowed us to use some GEO-lasercom-class technologies.

The Earth-based receiver was another thing altogether, though. Technologies for the detection of single photons had been known for some years, but most of them fell down in speed or size or the efficiency with which they actually detected incident photons. Furthermore, none of them had the kind of reliability or engineering ruggedness that would be required by a real communication system. Serendipitously for our study, Lincoln Laboratory had been developing a special kind of semiconductor detector, called an avalanche photodiode (APD) that, when combined into an array and run in what is known as the Geiger mode (GM), can be turned into a very efficient and adequately fast photon counter.

These Geiger-mode APDs had been first developed at Lincoln Laboratory for laser radar systems. A member of our study team knew about this work, and our system designers then met with the device designers. After several discussions, we all realized that an array of these GM-APDs could, indeed, serve as the basis for extremely efficient deep-space lasercom. And quite happily, the available speeds of the developed detectors perfectly matched the data rates we were seeking.

Some major developments were still required, however. Communication systems are different from laser radar systems in the patterns in which the pulses arrive. Our use of these devices would require a much more complex set of electronics designed into the integrated circuit onto which the detector array was bonded. In a quite amazing tour de force, a Lincoln Laboratory team took only about one year to plan out, design, build, and demonstrate a new integrated circuit to perform the required functions.

13.7.3 The Telescope Solution

We were almost there. We had concepts for the complete space system, modulation and coding approaches for the signal, and a receiver approach for the ground system. If we could find an Earth-based telescope system that could be used with these photon-counting detectors, and that could be used in turbulence, and that didn't cost too much, we would be in business.

It is known that there are three things one can do with respect to receiving light through turbulence.

- Build an adaptive optics system based either on the incoming light or on some natural (star?) or artificial beacon. Such a system potentially makes it look as if there is no turbulence.
- Live with the effects.
- Design the telescope to be small because it turns out the effects are less with a small telescope.

For the Mars scenario, not really enough light comes down to feed the sensors that an adaptive optics system requires. Building an artificial beacon would have been possible, but it would have added a lot of complexity, and we did not want to go that route for this first cut at a deep-space link.

We also knew that a large telescope incurs a large penalty from turbulence, and so living with the effects was probably not acceptable. Not to mention that large astronomical-grade telescopes are *very* expensive.

Radio engineers have gotten very good at creating arrays of antennas to do various special functions. Might it be possible to build an array of smallish telescopes, each with adequately small effects from turbulence, and then combine their outputs somehow? The answer was a resounding yes.

A photon-counting receiver takes light in and puts out electrical "ones." That is, it counts a one for each photon. If we knew the timing of each photon from multiple telescopes, we could "just" add them together digitally with the correct timing and create a new electrical digital signal that is equivalent to that which would have been captured with a huge photon counter behind a huge telescope.

We did an analysis to find the cost/performance trade-off between many smaller telescopes and a single large telescope. Of course, each system will have its own optimization. Some assumptions might lead to, say, four telescopes combined, and other assumptions might lead to 100 small telescopes. But the concept was valid and very powerful. One possibility, with 18 one-meter telescopes, is shown in Figure 13.10. This system has collecting power equivalent to that of a single 4.25-meter-diameter telescope, and is potentially somewhat less expensive. Furthermore, the effects of turbulence on each of these telescopes is quite a bit less than that on the large monolithic design.

Figure 13.10 Example telescope array.

So the solution we proposed was to use the most efficient class of detectors in a less expensive telescope array configuration that has smallish penalty resulting from turbulence.

Win, win, win—it is rare that engineering works out that way.

13.8 Good News—Then Bad News

We felt that we had solved the transmitted modulation + receiver architecture + telescope array + beam stabilization problems, as well as coming up with preliminary mechanical, electrical, optical, and thermal designs to enable all these things. We also were able to convince NASA that we were well on our way to successfully solving their original problem statement. And so after we briefed the results of our study, NASA immediately kicked off a major flight program—the Mars Laser Communications Demonstration—in which we, along with a team from JPL, were to implement these ideas and demonstrate 10s of Mbps (remember, this was the first time for trying to engineer many of these concepts) from Mars to Earth.

The program lasted several years and even passed a full set of Preliminary Design Reviews. Unfortunately, high-level governmental redirection of NASA goals led to the cancellation of the satellite our system was to fly on, and MLCD had to be cancelled as a result.

In the ensuing years, our group looked around and realized that expensive missions to Mars would likely be few and far between. However, we thought that missions to the Moon and possibly even the Sun-Earth Lagrange points (where, for example, the James Webb Space Telescope will reside) would be more likely, given their somewhat lower costs. Therefore, our group and other Lincoln Laboratory collaborators took the basic concepts we had developed for MLCD and started translating them into benefits for shorter, but still quite long, laser communication links.

The team took the hybrid tracking scheme and applied its basic concepts to an inertially stabilized small and gimbaled telescope, shown in Figure 13.11. Although not as critical for shorter distances, the inertially stable design allows systems to deliver very high downlink data rates while not having to work as hard on the uplinks since uplink rates will likely remain relatively low for almost all of NASA's near-term needs.

The team, along with others at MIT, developed a new photon-counting detector technology, called superconducting nanowires, which had several benefits: the ability to work at a preferred fiber telecommunications wavelength, much higher speeds (which can map into higher data rates that would become feasible at the shorter distances), and higher detection efficiency. Over several years, these new detectors were matured and then demonstrated in end-to-end high-rate lasercom demonstrations in the lab.

Finally, system architectures predicting substantial data-rate capabilities with already-developed space transmitter designs and proposed inexpensive ground arrays completed the picture. With these in hand, our NASA sponsors went looking for

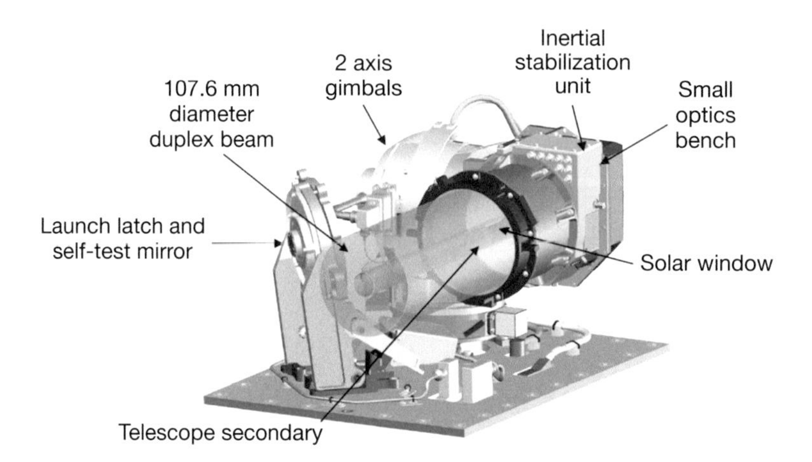

Figure 13.11 Small, inertially stabilized optical module for space-based laser communications.

a "ride" on which we could demonstrate this capability. After a few years, one such ride appeared.

13.9 More Than Half a Gigabit per Second From the Moon

In 2008, NASA's Science Mission Directorate announced that they were kicking off a mission called the Lunar Atmosphere and Dust Environment Explorer (LADEE). This mission was to be based on a novel, small spacecraft (Figure 13.12) that would be sent into lunar orbit and there perform measurements for several months of the very thin lunar atmosphere. The satellite had some reserve capability in terms of supportable mass and electrical power, so our sponsors in NASA's Human Exploration and Operations Mission Directorate enlisted us as a technology demonstration addition to the LADEE science mission. Our mission was named the Lunar Laser Communication Demonstration (LLCD) and consisted primarily of a space terminal to fly on LADEE and a "transportable" ground terminal that would reside at the NASA site in

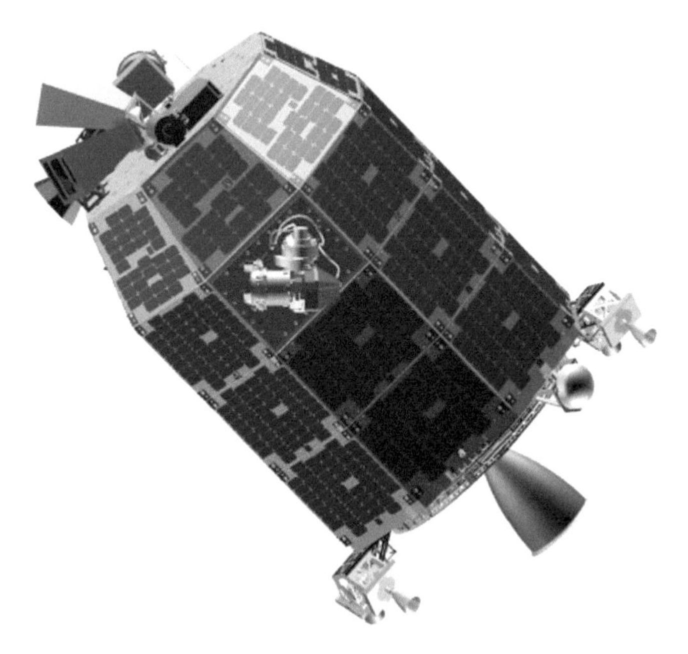

Figure 13.12 NASA's Lunar Atmosphere and Dust Environment Explorer (LADEE) spacecraft with the Lunar Laser Communication Demonstration space terminal's optical module visible in the center. (Rendering courtesy of NASA.)

White Sands, New Mexico, along with the overarching system architecture and design that flowed down to the detailed specifications for these two terminals. The space terminal was to be based on the detailed concept the group had been developing, and the ground terminal was to include a small ground telescope array plus the superconducting nanowires.

Between 2008 and 2012, the detailed designs were devised, refined, and analyzed, and the hardware was developed and "qualified" either for the space or ground environment. With the space terminal ultimately consisting of three distinct modules, each of which was built up of multiple submodules, numerous parallel teams were at work, coordinated by both the system engineering team and the program management team. Interfaces to the new little spacecraft also needed to be specified. With both the spacecraft and the lasercom system being designed at the same time, these interfaces were difficult, moving targets for both teams. With some technical and schedule give and take, though, from both sides, the space system design was finalized and had physically come together by early 2013.

The ground terminal was also developed by multiple parallel Lincoln Laboratory teams working on their separate modules—the cryogenic system including multiple nanowires, the high-speed processing electronics, the uplink transmitter electro-optics, the telescopes and gimbal, and the point-acquisition-track mechanisms and controls. Here, too, system engineering made sure that the modules not only worked together, but also that they would work with the distant space terminal. Many levels of functional and interface tests were run, with what we have come to call "rolling integration." With this philosophy, higher and higher levels of integrated testing can be done with whatever maturity version of each subsystem is ready at the time—breadboards sometimes, engineering units sometimes, and ultimately, the final versions. Such an approach allows the team to sniff out system problems early.

In fall 2012, the ground terminal was integrated at an outdoor site a short walk from Lincoln Laboratory, and it was put through its paces—everything but communicating with a distant terminal—during the winter and then spring 2013 months. Pointing, tracking, coupling of light into the receiver, and various calibrations were exercised using stars. In early summer, the transportable system was disassembled and shipped to White Sands, where it was reassembled and prepared for operations planned for fall 2013.

About a year and a half before the operations began, NASA had enlisted two other organizations to configure their already-existing telescopes for operations with LLCD: JPL with their telescope at Table Mountain in California, and the European Space Agency with their telescope on the island of Tenerife in the Spanish Canary Islands. The addition of these two other assets meant that there would be more hours per day with the Moon in view somewhere, and also a certain diversity to give more chances for operations if it happened to be cloudy at one or another site. Needless to say, it required even more careful interface testing between the Lincoln Laboratory–built space terminal simulator hardware and the new ground modems of each of these systems.

The rocket itself was yet another new design in the mission, a Minotaur V rocket, and it was launched from NASA's Wallops Flight Facility on the coast of Virginia in early September 2013. The night launch was visible all up and down the East Coast, and so got a lot of attention from the public (see Figure 13.13.)

A month after the launch, NASA's LADEE team successfully put the satellite into a two-hour lunar orbit. A schedule had been proposed whereby LLCD would get to operate on a seven-day cycle lasting one month, with four days for LLCD and then three for LADEE science preparations. Of course, LLCD could only operate when the Moon was in view, and so over the month, operations followed the Moon's appearance all across the 24-hour day and included all phases of the Moon.

Figure 13.13 Minotaur V with LADEE spacecraft, just before the September 2013 launch from Wallops Island. (Photograph courtesy of NASA.)

So… how did it work? Well, all the system design and analysis, and all the hardware design and testing details, and all the interface and integrated testing paid off. The system was a resounding success: acquiring and tracking the narrow uplink and downlink beams were instantaneous and robust in all conditions, the downlink communications signaling and receiver architecture provided error-free data delivery in nearly all cloud-free atmospheric conditions, uplinks were similarly robust, and various real data delivery was demonstrated, including the downlinking of the entire 1-gigabyte LADEE buffer in minutes instead of the multiple days it might have taken with the LADEE system. The system demonstrated the highest data rates of any communication system ever either from or to the Moon—622 Mbps down and 20 Mbps up—and demonstrated the longest free-space lasercom system ever by a factor of ten.

For a system engineer, it will be really tough ever to top this achievement. A number of people on the program have independently said this to me.

13.10 Lessons Learned

Other than the obvious technical successes of the LLCD mission with its various novel subsystems, did we learn anything nontechnical in the Mars study, the MLCD program, and then the LLCD project? Perhaps we can boil it down to two important lessons:

- Don't let a large gap between current practice and new goals stop your work. If there are known concepts that don't defy physics and that suggest the gap is not impossible to breach, you should aim high.
- Any new concept will have many unpredicted details that will all need to be discovered and then worked out before it can be fielded. But don't let those difficulties dissuade you from pursuing the ambitious concept.

About the Author

Don M. Boroson is an MIT Lincoln Laboratory Fellow. Prior to holding this position, he was the leader of the Optical Communications Technology Group. He has also served as assistant leader of the Satellite Communications Technology Group and as a member of the Senior Staff in the Communications Division. He was awarded a 2007 MIT Lincoln Laboratory Technical Excellence Award for his work on signaling for optical communications systems.

Dr. Boroson was the lead engineer on Lincoln Laboratory's GeoLITE project, the world's first successful high-data-rate, long-distance space-based lasercom system. He was also the lead system engineer on NASA's Mars Laser Communications Demonstration, and is currently the principal investigator and Lincoln Laboratory program manager for NASA's Lunar Laser Communication Demonstration. Dr. Boroson is widely published, is internationally known for his work in optical communication systems, and is a Fellow of the International Society for Optics and Photonics (SPIE). He holds BSE, MA, and PhD degrees in electrical engineering from Princeton University.

Summing Up

William P. Delaney

Is Systems Analysis for You?

Here, at the end of our book, I address a few points about the picture we have painted of systems analysis and speculate on the future prospects for the systems analysis process. After reiterating some of the common themes and perspectives of the authors, I finally ask readers, "Is this systems analysis stuff for you?"

Too Rosy a Picture?

By now, our readers surely sense that our authors "love" this systems analysis art, but I worry we may have created a bit too rosy a picture of the impact of one's systems analysis efforts. Our selected examples often point to some decisive outcome or at least a favorable outcome; this does happen, but not as frequently as one might hope. What does happen is a clearer understanding of the problem for yourself, your colleagues, and your sponsor. This understanding is often the first step in influencing a larger decision or the start of some national investigation. Along the way, you map out the problem in plain English for the leadership and advise them on the technical uncertainties of the issue in question. You provide support to researchers in government who are on the right course to solving the problem. Most of us have found that this kind of analysis needs to precede our involvement in any substantial project. So systems analysis is a great discipline even if it does not hit a home run every time.

Systems Analysis in the Future

I argue that the type of systems analysis we have explored will continue to be important in the future. Technology will provide more and more powerful tools in sensing, computation, automation, robotics, and biological systems, but the application of these tools to vexing problems of today and the future will remain a task for the inquisitive, broad-thinking mind. Systems analysis will move increasingly outside the defense arena and be applied to issues such as global warming and sustainable energy. The technology and the numbers really do count in these global challenges. David

JC MacKay in his book[1] on sustainable energy provides a wonderful systems analysis example of the energy situation, complete with technology and numbers, numbers, and more numbers. He analyzes the options to migrate Great Britain to a totally "green" energy state, and along the way he lets a lot of "hot air" out of the energy debate. I am convinced that these global issues plus continuing national security issues will guarantee that systems analysis is here for the long pull.

Common Themes

We probably hammered to death some themes in the past several hundred pages:

- Get the question right at the start.
- Do not bog down in details. (However, some details really count!)
- Back-check your work with simple calculations and a sanity check by asking yourself, "Does it make sense?"
- Be suspicious of your assumptions and keep them transparent to your audience.
- Pursue the elusive "truth" in your analysis and your presentation.
- Use experimentation (lab, field, whatever is available), if possible, to check your results and to resolve technical unknowns.
- Be cautious with computer models and simulations—the more complex the model, the more cautious.
- Tell a good story—interesting, logical, focused on the "why," and explaining why it makes a difference—and make it a short story!
- Bring along new folks and introduce them to the art.

Is This Stuff for Me?

I said in my very early remarks in the foreword to this book: "… a technical organization does not need a lot of systems analysts, but it surely needs a few." The newcomer to this type of work is likely asking, "Is systems analysis a good career path for me?" We have probably made it sound pretty good. Let me review some upsides and downsides of this type of work.

1 D. JC MacKay, *Sustainable Energy — without the hot air*. Cambridge, U.K.: UIT Cambridge, 2009.

Upsides

- You engage the problem quite early and have an opportunity to shape both the ongoing work and the future. This is an exciting time, and your work can be high leverage and on the cutting edge.
- Your work is aimed at the high-level decision makers, and it is of high interest to them.
- You get to interact with some of the "best and the finest" in the science and engineering community. Systems analysts are often the intellectual leading edge of a big organization, and they usually are pretty sharp folks with broad perspectives.
- Your work is the early phase in a large program, and you likely do not get heavily involved in follow-on activity with its technical challenges, shortfalls, schedule slippages, funding cuts, and all the other agonies that seem to accompany a big engineering project. Either you will find this an upside or a downside; if the latter, systems analysis is not for you.
- You get exposed to lots of different problems at a fairly rapid rate, so there is not much time to get bored.
- You get to tell interesting and captivating stories as you take a complex problem apart and then put it back together in an understandable form. You will often find that after disassembling and reassembling a complex problem and turning it into a good story, you understand the issues better than 95% of all the other people who are in the loop.
- You have the pleasure of introducing some new bright folks to the systems analysis art and watch them grow and, if you really do this well, surpass you!
- Finally, the broad-thinking of the systems analysis art can be helpful in evaluating courses of action in everyday life (e.g., rent versus buy, planning for retirement, etc.)

Downsides

- You don't get to "bend" hardware, but it is not a bad idea to have started out your systems analysis career with some time spent building things; it will give you a very useful and important perspective.
- You don't get to do deep dives in some fascinating technical area.

- Over a long career, you do not leave many "monuments" to yourself (big radars, new aircraft, new computers, advanced ships, etc.) behind on the landscape. But on the other hand, you will be able to take some pleasure in knowing that those monuments look the way they do because of your early involvement.
- You probably won't form your own company to produce some widget you've invented and thereby get rich! (Conversely, you can form an analysis company in the "beltway bandit" style and get very well off, if not rich.)

Final Thoughts

Most of our 10 authors have had careers that are not 100% systems analysis but have at times involved hardware or software engineering and management of large projects. I, for one, feel I have left a few proud monuments on the landscape. The systems analysis art and perspective served me well in these big projects. I paid a lot of attention to "Why this project?" and "How can we do it?" and I explained those aspects of the project to both the team that was developing the system and the outside world that was being impacted by our great new "stuff."

My bottom-line message to our readers is that this protocol of taking a very broad view of a scientific or engineering challenge, understanding how the big parts fit together, and always asking yourself "Why?" is a nationally important discipline and a wonderfully useful discipline whether you're a high-profile advisor to leadership or someone down in the trenches trying to make an engineering vision a reality.

Good luck to all!

Bill Delaney

At the close
Summer 2014
Lexington, Massachusetts

Acronyms

AADS-70 Army Air Defense System – 1970
ABMDA Advanced Ballistic Missile Defense Agency
ADSEC Air Defense Systems Engineering Committee
APD avalanche photodiode
ARPA Advanced Research Projects Agency, now DARPA
ARSR Air Route Surveillance Radar
ASAT anti-satellite system
ASB Army Science Board
ASD(R&E) Assistant Secretary of Defense for Research and Engineering
ATBM advanced tactical ballistic missile
ATR automatic target recognition
BMC^3 battle management, command, control, and communications
BMD ballistic missile defense
BMDS Ballistic Missile Defense System
BMEWS Ballistic Missile Early Warning System
C2 command and control
C3 command, control, and communications
C3I command, control, communications, and intelligence
CDC Centers for Disease Control and Prevention
CERN the European Organization for Nuclear Research
CONOPS concept of operations
CONUS continental United States
CPU central processing unit
CSA cyber situational awareness
DARPA Defense Advanced Research Projects Agency, formerly ARPA
DCA defensive counter air
DDR&E Director of Defense Research & Engineering, now ASD (R&E)
DHS Department of Homeland Security
DIK detect, identify, and kill
DMSP Defense Meteorological Satellite Program
DoD Department of Defense
DSB Defense Science Board
DSC defensive space control (see also OSC)
DSN Deep Space Network
DSP Defense Support Program
DTRA Defense Threat Reduction Agency
EP electronic protection

EPA	Environmental Protection Agency
ERP	effective (sometimes equivalent) radiated power
FABMDS	Field Army Ballistic Missile Defense System
FEBA	forward edge of battle area
FEZ	fighter engagement zone
FFRDC	federally funded research and development center
FLOT	forward line of own troops
GM	Geiger mode
GMTI	ground moving target indicator
GPS	Global Positioning System
HSD	hard site defense
HTK	hit to kill
HVAC	heating, ventilation, and air conditioning
IADS	integrated air defense systems
ICBM	intercontinental ballistic missile
ICD	interface control document
IFF	identification, friend or foe
IFOV	instantaneous field of view
IOC	initial operational capability
IPR	interim program review
IR	infrared
IR&D	independent research and development
IRU	inertial reference unit
IT&E	integration, test, and evaluation
JPL	Jet Propulsion Laboratory
JSTARS	Joint Surveillance and Target Attack Radar System
LADEE	Lunar Atmosphere and Dust Environment Explorer
LCC	life-cycle cost
MDS	Minuteman Defense Study
MIPS	million instructions per second
MTI	moving target indication (see also GMTI)
MWRA	Massachusetts Water Resource Authority
NASA	National Aeronautics and Space Administration
NATO	North Atlantic Treaty Organization
NRAC	Naval Research Advisory Committee
OMB	Office of Management and Budget
OSC	offensive space control (see also DSC)
PAG	power-aperture-gain
PCR	polymerase chain reaction

PMO	project management office
PRF	pulse-repetition frequency
QVN	identification label of the ARSR-3 located at Fossil, Oregon
R&D	research and development
RCS	radar cross section
RF	radio frequency
RV	reentry vehicle
SAM	surface-to-air missile
SAM-D	surface-to-air missile, development
SAMSO	Space and Missile Systems Organization
SAR	synthetic aperture radar
SDI	Strategic Defense Initiative
SLS	shoot-look-shoot
SOSI	space-object surveillance and identification
SPIE	International Society for Optics and Photonics
SSA	space situational awareness
TBM	tactical ballistic missile, also tunnel boring machine
TMD	theater missile defense
TOR	Terms of Reference
TRL	Technology Readiness Level
UAV	unmanned aerial vehicle
USGS	United States Geological Survey
VIN	vehicle identification number
VIRADE	virtual radar defense, developed by Bell Laboratories
WMD	weapons of mass destruction
WP	Warsaw Pact

Printed in the United States
by Baker & Taylor Publisher Services